D0723547

Transformations

New Studies in Phenomenology and Hermeneutics

KENNETH MALY
Series Editor

193
H465M
S826t

22.46

Transformations

Thinking after Heidegger

GAIL STENSTAD

THE UNIVERSITY OF WISCONSIN PRESS

The University of Wisconsin Press
1930 Monroe Street
Madison, Wisconsin 53711

www.wisc.edu/wisconsinpress/

3 Henrietta Street
London WC2E 8LU, England

Copyright © 2006
The Board of Regents of the University of Wisconsin System
All rights reserved

5   4   3   2   1

Printed in the United States of America

Library of Congress Cataloging-in-Publication Data

Stenstad, Gail.
Transformations thinking after Heidegger / Gail Stenstad.
p. cm.—(New studies in phenomenology and hermeneutics)
Includes bibliographical references and index.
ISBN 0-299-21540-7 (cloth: alk. paper)—
ISBN 0-299-21544-X (pbk.: alk. paper)
1. Heidegger, Martin, 1889–1976. 2. Thought and thinking. I. Title. II. Series.
B3279.H49S696 2005
193—dc22
2005005467

IN MEMORY OF

BILL BOB,

For the sake of all living things:
May we wake up and listen to them.

To think is above all else to listen.

—MARTIN HEIDEGGER

CONTENTS

PREFACE

This book is for anyone who has ever read Martin Heidegger and thought, "This is interesting, but what use is it?" In particular, this book is for anyone who has read the later works of Heidegger and caught a glimpse of something significant but elusive. We read in Heidegger about technology and about our destructive domination of the earth's resources. We read about how we exploit each other and about our use of language as mindless entertainment or, worse yet, destructive propaganda. We read about our general failure to think for ourselves and other issues that are of concern to many of us. At the same time, if we are careful readers, we can see that Heidegger offers no ready solutions and even deliberately renounces the usual paths to such solutions, namely, theory, value judgments, and ethics. Furthermore, what Heidegger says is often rather difficult to understand, even for scholars. What are we to do with such a path of thought? I hope to offer some clear indications of a response to that question. This book will, I hope, be of interest to Heidegger scholars and will generate discussion of the possibly controversial suggestions I am going to make. However, I think what Heidegger has to say is too important to remain buried in the halls of academia.

In *Transformations* I explain what are generally considered to be some of the most difficult matters in Heidegger: be-ing as ab-ground, timing-spacing, emptiness and opening, and his utter refusal to give in to any demand to theorize or moralize. However, even though I hope to challenge Heidegger scholars to think more carefully and creatively and openly, I am writing for a broader audience. At the heart of my book is the work of carefully and persistently clarifying the intimate relationship of deep thinking, everyday life, and the "big questions." Rather than just talking about Heidegger's work of thought, I have taken up some of his keys to *doing the work of thinking*, and I use them to lead the reader along with me in engaging in thinking about some of the vital issues that were at stake for Heidegger. They are still very much at stake for us in the twenty-first century. I aim to call the reader, whether inside or outside academia, to engage with me in enacting a dynamic *way*—not a theory, not a method— of thinking that can, quite possibly, transform our destructive and exploitive relationships with nature, our fellow living beings, and one another. Whether we are concerned primarily with environmental issues, sexism and other forms of exploitive dominance, or the ever-increasing tendency toward reactive violence in our world today, we struggle with the dissonance between the need for change and our apparent inability to make those necessary changes. Thinking with *and after* Heidegger opens ways to transformatively confront this dilemma in whatever areas most deeply concern us.

I thank Kenneth Maly, LaDelle McWhorter, and Patricia Huntington for their thoughtful and helpful comments, criticism, and encouragement. I am especially grateful to my husband, James Oler, for having read every draft of every chapter with pen in hand. His suggestions, both editorial and substantive, have made this a much better book.

ABBREVIATIONS

of Selected Works by

MARTIN HEIDEGGER

BT	*Being and Time*
BW	*Basic Writings*
CP	*Contributions to Philosophy (From Enowning)*
DT	*Discourse on Thinking*
EGT	*Early Greek Thinking: The Dawn of Western Philosophy*
GA 2	*Sein und Zeit. Gesamtausgabe*, vol. 2
GA 5	*Holzwege. Gesamtausgabe*, vol. 5
GA 7	*Vorträge und Aufsätze. Gesamtausgabe*, vol. 7
GA 9	*Wegmarken. Gesamtausgabe*, vol. 9
GA 12	*Unterwegs zur Sprache. Gesamtausgabe*, vol. 12
GA 45	*Grundfragen der Philosophie. Gesamtausgabe*, vol. 45
GA 65	*Beiträge zur Philosophie (Vom Ereignis). Gesamtausgabe*, vol. 65
ID	*Identity and Difference*
P	*Pathmarks*
PLT	*Poetry, Language, Thought*
QT	*The Question Concerning Technology and Other Essays*
TB	*Time and Being*
WCT	*What Is Called Thinking?*
WHD	*Was Heißt Denken?*
WL	*On the Way to Language*

Transformations

Introduction

Many years ago, when I lived in rural northeastern Iowa, something happened that—though I didn't fully realize it at the time—would serve as a touchstone for the course that my thinking would take much later in higher education and beyond. It was an ordinary day, much like any other summer afternoon. I had been doing a little light weeding in the vegetable garden and was standing on the lawn nearby, admiring the results. I wasn't feeling especially tired, or excited, or anything other than just calm and content with the beauty that I saw as a result of my work. Then, without any verbalized intervening thoughts, I lay face down on the grass and (I know this will sound very strange), quite simply, I was the earth. This was no vague at-one "feeling." It was very clear, very precise. I did not lose awareness of my body, engaging the solidity of the earth with its own density. But neither was that meeting of earth and bodily awareness a barrier. I simply *was* the earth, and my awareness was the earth's awareness, moving through space, with the stars glittering in the blackness of surrounding space.

I suppose, things being the way they are in our world, that I should add that I was a straitlaced young woman who had never experimented with drugs. And, as I was then a Midwestern Protestant Christian, I had not been engaging in any meditative practices that could have opened me to something like this or helped me to understand it. *It just happened*, spontaneously, seemingly without context and without precedent. The lack of precedent is not entirely true—I had had a spontaneous and unquestioning closeness to the earth as a young child. But with that awareness having

receded into the deep, nearly forgotten background, and with no explanatory context, I did not know how to make sense of this. Oddly, perhaps, I
had no doubt whatsoever about the validity and importance of what happened, but I did keep it to myself.

That something significant and radically different was going on was
borne out a couple of days later. Again, I was outside, alone. I don't recall
what I was doing. I knew there was a family of woodchucks living under
one of the outbuildings, as I had caught quick, occasional glimpses of them.
What happened now was that as I was standing in the front yard the adult
female woodchuck came around the corner of the house. I looked at her;
she looked at me. I went down on my haunches to be lower and less intim-
idating and began a soft, monotonous whistling (that was odd in itself,
since I am rather whistle-impaired). She came forward, slowly and quietly, put one of her front paws on my foot, and looked up at me. We stayed
there, quietly together, for some time, though I have no idea how it would
be measured on a clock. Then she turned around, went back around the
corner, and that was that. It never happened again. It would be a good
long while before I found out that there was a way to *think* about such
things without just rationalizing and explaining them away.

Life went on, with its busy activities and its ups and downs. A few years
later I went back to school, going to our local regional university with the
intention of majoring in biology. In my second year, however, I took a couple of philosophy classes and was hooked, particularly by the course in
ancient philosophy, which was taught by someone who not only knew the
Greek but taught the course (though I didn't know this until later) from
the perspective of the kind of thinking about philosophy's history that had
been opened up by Heidegger. Amazing things: Antigone, Socrates, Anaximander and Heraclitus and Parmenides, Plato! Even more amazing was
the notion that there was more than just the surface meaning. Besides what
we could readily see because it fit with all our received presuppositions,
there were also the things *unthought*; there were hidden possibilities that
were only intimated but that could be drawn out and held in a carefully
questioning thinking process. More amazing yet was that this way of thinking was not something that could be "learned" in the usual way, to be spit
back on some multiple-choice test or even on some "objectively graded"
essay. You had to actually *do the thinking yourself*, and what you would think
could not be predicted ahead of time, and *that was quite all right!* In the
next philosophy course I took we read Heidegger's *Discourse on Thinking*,
and I knew now at least one source for this notion of the unthought and of

a radically different way to think. So I started reading Heidegger, not *Being and Time* but the later works, especially "Building Dwelling Thinking," "The Thing," and *On the Way to Language.* When I got to graduate school and took a Heidegger seminar, the text was *Being and Time.* To be quite honest, the phenomenological descriptions of Dasein left me indifferent, but I zeroed in on sections 1–7, 34, and 68: the question of the meaning of being, the use of the hidden meanings in the Greek to unfold a different, deeper, and more open sense of the meaning of phenomenology, and the place of language in both opening and limiting Dasein's world. Back then (around 1986) I thought it was odd that the book never quite got around to actually saying anything much about time, beyond (1) Dasein's temporality in its holding open the place for the disclosure of the significance of the beings in the world and (2) the leveling off of temporality into public, measurable clock time.

I was on the right track for at least two good reasons. "In the question of being, we are dealing solely with the enactment of this preparation for our history. All specific 'contents' and 'opinions' and 'pathways' of the first attempt in *Being and Time* are incidental and can disappear. But reaching into the free-play of time-space of be-ing must continue" (GA 65: 242/CP 171). So there is much more to be thought, and this thinking has still only just begun. The way of thinking opened up by Heidegger is a powerful way—though by no means the only way—to recover much that has been lost to us through the dominance of a way of thinking and living that disallows genuine questioning and, in so doing, disempowers us and is on the brink of destroying the earth and our fellow living beings. Recovering lost possibilities is already something significant, but I think if we can follow through with thinking after Heidegger, we will also open the way to as yet unimagined possibilities. This is no panacea; it takes work and the willingness to be radically changed. But, I wager, the effort and risk are well worth it.

Before I give a preview of each chapter I want to point out three things about the structure of the book. First, chapters 1 and 2 contain things that are, for the most part, fairly well known to Heidegger scholars, though I am going to make connections between thoughts that I have not seen elsewhere. Chapter 1 is, in its appearance, perhaps the most "scholarly" of the chapters, holding most tightly to the texts under consideration. In the remaining chapters I open the thinking up in ways that I hope will challenge and provoke the scholars to see Heidegger's work in a fresh light. Second, chapters 1, 3, and 5 focus on a careful and close reading and interpretation of certain "themes" in Heidegger's thinking (the question of the meaning of being, time-space, and emptiness), while chapters 2, 4, and 6 focus on the kind of thinking that is called for and the relationship of thinking (and what calls for thought) with dwelling, with our ways of living in this world. I mention this distinction only because I think it may be helpful to the reader, but I also want to caution against making too much of it. The specific contents or themes are, in fact, inseparable from the way of thinking they require, and dwelling *spontaneously* arises from doing the thinking. The nature of this inseparability, however, will only emerge in the course of the thinking. It begins to appear in the language that carries the thinking, for instance, in the repetitions that necessarily come into play. While some of my use of repetition in writing this book is simply recapitulation and review to help the reader follow the train of thought, much of it does something else. Each time something is repeated, unless it is in an obviously straightforward recapitulation, the repetition occurs in a context in which the meaning of what is repeated shifts. This shifting itself both deepens the understanding of that particular item and shows something significant about how this thinking unfolds transformatively.

In addition to a distinction between the matter to be thought (chapters 1, 3, and 5) and the way of thinking there is another movement in the structure of the book, the third structural feature to be pointed out. The first three chapters begin to unfold—through both explaining and enacting—what it means to understand thinking as a *way*, from the question of the meaning of being, deeper and deeper, all the way into timing-spacing-thinging. Chapters 4 through 6 increasingly show what is required of us—what demeanor, what manner of engagement—if we are to think and to let that thinking matter in our lives. These chapters focus on our understanding of ourselves and the things of the earth and the world as well as the nature of the relationships we live and enact with them. Again, this structural distinction is not rigid. For example, the "releasement toward things" discussed in chapter 2 clears away key hindrances to thinking and is thus a necessary part of our demeanor in the face of the call to think. With that said, let me give a brief overview of the focus of each chapter.

Chapter 1 relates just how it is that Heidegger opens a way into this powerfully transformative thinking in posing and attempting to come to grips with the question of the meaning of being, a question that requires thinking historically. This is not just an abstract discussion of the history of metaphysics. Heidegger's thinking points to the ways in which that history still holds power, and he connects it to issues of deep concern to many: the drive for maximum organization and control, environmental devastation, and humans' use and abuse of one another. Heidegger helps us to see how our idea of "being" arose in the creative thinking of the ancient Greeks in response to their wonder at the arising into presence of beings. They began to think of "being" as something distinct from "beings" and, not noticing the creative nature of their own thinking, accepted this concept of being in the light of a discovery. From then on, as Western metaphysics unfolds from that beginning, various interpretations of being as the ground of beings determine our ways of relating to those beings. Understanding that, we can also understand how this history shapes us now, even in terms of our relationship to modern technology. Heidegger also—and this is crucial—leads us to see that in thinking this beginning of Western philosophy we are *already* engaged in opening the way to a new beginning, as the notion of the reification of being comes in for serious questioning. Instead of "ground," we begin to explore ab-ground, the staying away of all such grounds, which also shifts and loosens the grip of being. Heidegger begins to mark this shift by the use of the word "be-ing" to emphasize its dynamic and questionable character.

Chapter 2 aims to help us reflect on how to hold be-ing in question and to *think* its ramifications. What sort of language is called for in this thinking? How does thinking open and move? One of Heidegger's suggestions is that we approach the matter through releasement toward things and openness to mystery. To even begin it is necessary to release old assumptions about the nature of thinking and about the language that carries it, because they arose along with the centuries-long development of interpretations of being and, as such, can only hinder our attempt to think be-ing, which is radically different and not subject to capture in the same kind of language. The nonreifiability of be-ing thus requires us to engage the matter with nonreifying language. But what can that possibly mean? To begin to answer that question and open up a way to proceed is the main topic of chapter 2.

If beings are not "beings," how are we to think them? That is the guiding question for chapter 3. Heidegger takes up the most ordinary of words—"thing"—and gives us a way to begin to think some of the most difficult questions in all of philosophy by way of it. Things are not "beings" but rather "thinging," which is the dynamically relational gathering of many as one but in such a way that the distinction of "one and many" itself comes in for questioning. To think very far into the question of the meaning of the thinging of the thing also calls for inquiring into the nature of the time and space in which it takes place. The entire discussion of timing-spacing-thinging

strongly reinforces the nonreifiability of the matter for thinking, be-ing.
Be-ing is not "being" but rather timing-spacing-thinging, in which we
also arise. This first raises the difficult but necessary question of our own
reifiability.

The farther we take that thought, on into chapter 4, the harder it is to
evade the insight that the description of timing-spacing-thinging includes
us as more than just an observer or a bit player. We, too, emerge in the
dynamic relationality of timing-spacing-thinging. This insight indicates
that along with the demise of old assumptions about being and beings
comes the deep questionability of our long-cherished, strikingly dualistic
notions about ourselves: rational animal, subject in a world of objects, and
mind in a body, distinct from all other bodies. But, then, how are we to
think of ourselves? My first attempt to respond to that question shows that
we do not yet have a full enough understanding of the nature of thinking
to tackle it, unless we realize that it is not just "mind" that thinks. Here I
will call on the oddly ignored "thanc," the heart-mind, introduced by Heidegger in *What Is Called Thinking?* to help us break through this barrier.
I also undertake to give some down-to-earth indications of what it might
mean to bring heart-mind to bear on our relationships with things and,
as Heidegger puts it, to think for the sake of dwelling.

My growing emphasis on the importance of letting go of our longstand-
ing compulsion to reify everything has to be directly confronted at some
point. Heidegger, particularly in *Contributions to Philosophy (From Enowning)*, speaks of be-ing (timing-spacing-thinging) using words like the "not"
in being, nothingness, nihilating, and emptiness. Yet he also makes many
cautionary comments against taking any of this to a nihilistic extreme. How
are we to understand this? And, even more important, how does it help
us to *think* so as to *dwell* with things? These are central questions for chapter 5. That chapter, which brings Buddhist thinking into play with these
questions, also opens up a different kind of pathway into dialogue between
philosophical and spiritual traditions, a way not tied to "comparative" philosophy or religious studies. Emptiness, we shall see, is no mere abstraction
but is another way to think and say timing-spacing-thinging and to better
understand ourselves and our relationships with things.

The main task of chapter 6 is to gather all the threads together, the
pathways (thinking, language) and what they open onto (be-ing, timing-
spacing-thinging, emptiness, dwelling with things). In this gathering it is
opening and *staying* open that come most powerfully into play. Releasement
toward things (letting go of various hindrances to thinking) and openness

to mystery converge in enabling a knowing awareness of ourselves as timing-spacing-thinging. But what does that mean day to day? The key to beginning to understand how we might each, one by one, respond to that question lies in thinking and enacting *opening*. And it is this opening that makes possible the radical and manifold transformations that may arise in thinking with and after Heidegger and that may enable us to free ourselves from dualistic, violent reification of self, others, nature, and things.

And what about the woodchuck, coming to encounter me face to face? Her eyes and mine, all four of them the eyes of the earth. She had something to say, if I would have known how to listen. All these years I was learning how to listen, which, as Heidegger says, is learning how to think. Thinking is learned only by doing it. I invite the reader to come along with me as I do some thinking and risk its transformations.

1

Opening a Way

The Question of Being

"Do we in our time have an answer to the question of what we really mean by the word 'being'? Not at all. So it is fitting that we raise anew the question of the meaning of being. . . . Our provisional aim is the interpretation of *time* as the possible horizon for any understanding whatsoever of being" (GA 2: 1/BT 21).[1] Every student or serious reader of Heidegger, from undergraduates studying the material in *Basic Writings,* to graduate students and scholars, as well as many among the well-read general public, knows the beginning of *Being and Time.* What is not so well known or understood is what this reawakening of the question of the meaning of being is supposed to be *for* and what becomes of the "question of being" after *Being and Time* rather abruptly ends without ever arriving at its provisional aim of the interpretation of time as the horizon for being. *Being and Time* was never meant to be a theory of human nature, as in the mistaken "existentialist" interpretation of Heidegger. "*Being and Time* is the crossing to the leap (asking the grounding question). As long as one accounts for this attempt as 'philosophy of existence,' everything remains uncomprehended" (GA 65: 233/CP 165). The key is, as Heidegger later tells us, that *Being and Time* opens up a dynamic path of thinking on which *everything is transformed.* Everything. *Being and Time* is, for Heidegger, the necessary opening into a multifaceted region of transformative possibility (GA 65: 475/CP 334).

Throughout Heidegger's long career as a teacher and writer nearly everything he spoke on and wrote touched on the matter of transformation, either lightly and almost as a reminder or explicitly and at length. In fact, it is rather remarkable, when stepping back to get a broad view of the material, that so many of the "key texts" of Heidegger not only do this but also touch on the particular areas of thinking or acting that are at stake for transformation: language, thinking, traditional ideas of "being," "truth," earth and world, time and space, openness and opening, and our day-to-day living with technology and its way of shaping our thoughts and actions. Furthermore, it is not that you will find a lecture here that discusses technology and an essay over there that deals with language and a book farther down the shelf that delves deeply into the nature of thinking. We do, of course, find those things (in "The Question Concerning Technology," "The Way to Language," and *What Is Called Thinking?*), but each of these works does much more. In looking carefully at many of the texts I consulted in writing this book, I noticed something significant in terms of their contents. Not counting *Being and Time* and *Contributions to Philosophy (From Enowning)*, which are long and comprehensive books fundamental to Heidegger's entire long work of thinking (and thus would be expected to contain references to all the "key issues"), I looked at the fourteen other texts I would be referring to most often.[2] Of the fourteen, each of which had its own particular focus, I found that most of them also discussed not just one or two of the other issues but several. Twelve discuss language, transformation, the dominance of techno-calculative thinking, and either space, time, or time-space. Eleven mention opening (or the closely linked clearing or lighting), thinking, and "being" as it is thought metaphysically. Ten explicitly refer to overcoming the subject-object distinction, and to enowning (*Ereignis*). A good half of them discuss earth (and world) and the nature of "the thing." This is even more remarkable in that only one of the fourteen texts is of book length.

What does this mean? Why is it significant? Heidegger's "one question" has many facets. In a way, that comment seems almost trivial. We take "being" to refer to anything that exists in any way at all. So, of course, that includes *everything!* That seemingly trivial obviousness, however, only covers over what is really going on here. Because being "includes everything," a transformation in the way we think "being" is going to bring a change in, at the very least, thinking and language (the "is"). But thinking and language shape our understanding of time, space, things, and ourselves. In the contemporary world our understanding of all of these things is also shaped by science and technology. So it is not just the fact that Heidegger questions the meaning of being that results in his ongoing concern with all these other areas but that his way of thinking and questioning concerning being is already in and of itself *radically transformative*. One of the ways it transforms thinking is in the direction of a much clearer idea of the *dynamic relationality* of everything that is. At the moment, this early on, I can only assert this: change one key thing, and everything else changes, too. For Western philosophy the "meaning of being" is the keystone. Move it, change it, and everything else changes; remove it, and the whole metaphysical edifice falls. And, as will gradually become more and more clear, all these crucial matters (being, time, space, language, thinking, mind, technology's dominance, things, earth, world, us) resonate and dance with one another in a complex and dynamic intertwining. Again: change one thing, change everything. Not, however, in the way we usually think of change, in terms of linear cause and effect. How, then? This must emerge as we go on.

Conceivably, we could enter this web of gems at any point, any facet. As Heidegger said more than once, genuine thinking is not just following a track predetermined by someone else (whether the bland but powerful "they" of *Being and Time* or a great thinker like Heidegger himself); rather, "it is enough if we dwell on what lies close and meditate on what is closest; upon that which concerns us, each one of us, here and now" (DT 47). In the long run that becomes *necessary*, unless this work of thinking is Heidegger's and his alone. But to jump immediately there would perhaps be premature. We are so strongly shaped and constrained by our linguistic, intellectual, and cultural inheritance that "thinking" and "thinking" are not the same! That is, we think we are thinking, but we may well be running along in the same old rut, the gerbil-wheel that society gives us to play with in our cages. This is not, on the other hand, to assert that "following Heidegger" is the only way to go. In fact, following as in copying or repeating or shuffling apparent facts and propositions into some kind of cohesive theoretical structure is precisely *not* the way to go. In his most extended discussion of the nature of thinking Heidegger told the students attending the lecture course that if that is all any of them wanted to do, "in that case, burn your lecture notes, however precise they may be—and the sooner the better" (WHD 160; WCT 158). There are other ways into transformative thinking, starting from other questions. In chapter 5, for instance, I bring the Tibetan thinker Longchenpa into dialogue with this thinking; he has a different starting point but thinks into a similar region of transformation. But the work of thinking that we call "Heidegger" is deep and powerful and filled with openings into thinking that—and this is vitally important—start from where we are now, in the contemporary Western world. It is, therefore, an excellent and perhaps even necessary place to start.

The way that will be taken here is, first, to situate Heidegger's one question (and its first major elaboration in *Being and Time*) in the context of transformative thinking or, as it is called in one of its major arenas, *Contributions to Philosophy (From Enowning)*, "being–historical thinking." So I discuss the historically situated transformation of thinking concerning "being" as laid out in that book first and then return to *Being and Time* to examine it in the light of its place in that region of thinking.

The First and Other Beginning: A Preliminary Sketch

Contributions to Philosophy (From Enowning) (which I will usually refer to simply as *Contributions*) was held back by Heidegger and not published until several years after his death (it was written during 1936–38 but only published in 1988). Why? Heidegger himself said that the book did not have the finished academic form that would constitute a publishable work. In hindsight we also see that Heidegger meant to first give us a context within which this work could be understood, particularly, the many lectures in the history of philosophy. It is a record of creative thinking in progress. By that I do not mean it is a lengthy series of notes that could have been revised into an academic book. "Thinking in progress" here means *thinking*, which is always on the way and brings to language what arises on the unfolding, shifting path of thinking. Chapter 2 discusses this on-the-way nature of thinking in much more detail. For now, we need to be aware that in *Contributions*, perhaps more than in anything else he wrote, Heidegger made an *extended* attempt for the language to show not just the thoughts but at the same time the dynamic "structure" of the thinking. The book is made up not of chapters but of "joinings." What is the significance of this? Books ordinarily have chapters ordered and numbered in sequence. Of course, due to the nature of written language, the joinings of *Contributions* have this apparent sequence, but they do not constitute a step-by-step, ordered, and systematic discussion of the apparently given theme (enowning). The book does not follow any philosophical method, nor does it construct a system. On the other hand, the joinings are not just a series of random epigraphs. There is something altogether *other* going on here. The joinings echo (in fact, "Echo" is the title of one of them), mirror, and play forth (the title of another) with each other as they resonate within the dance of thinking. Any one of them opens ways into all the others, like many multifaceted jewels, each of which reflects all the others. To extend the image, they are not jewels set in solid silver but jewels hanging on a very large but delicate wind chime, responding to each nuance of the breeze of thinking. Each of the nonsequential joinings says the matter in a different way, with a somewhat different bearing, but at the same time they resonate as a conjoined questioning-thinking, "saying the same about the same," in the sense that that phrase carries in Heidegger's work. Though there is some repetition, the joinings do not merely repeat each other or say something identical; they say what intrinsically belongs together within the same region of thinking, moving -thinking along through the resonance of the nuances of their differing ways of bringing the matter to language (GA 65: 82/CP 57).[3]

What is the place or role of *Contributions* in the large work of thinking that we call by the name of Heidegger? It is certainly not a hidden system that organizes all the later lectures and published works. We could say, though, that it serves as a *touchstone* for them. It is helpful to think of Heidegger's later lectures and publications as joinings, too, interrelating with *Contributions* and with each other in a manner similar to the way that the joinings of *Contributions* function internally. As I have already pointed out, most of Heidegger's works at least touch on all of the key "themes" to be taken up by thinking, just as the joinings of *Contributions* all echo different facets of one overarching matter: the possible transformation of thinking in the leap into the interplay of the first and other beginning of Western philosophy. The different works of Heidegger, whether they are in dialogue with the great thinkers of the tradition (which will not be a focus of this book) or are among the fourteen I mentioned above, can be seen as joinings with each other and with *Contributions*, with *Contributions* serving as a touchstone but not the last word. That is, at times the others carry thinking farther than does *Contributions*. Well, of course! It would be very odd if Heidegger's thinking stagnated after 1938 and all he did was mine *Contributions* for publications and lecture material. However, the main thing to be aware of is that *all* of Heidegger's lectures, published essays, and books are "thinking in the crossing" of the first and other beginning, with some of them having more of an obviously preparatory character and others being more clearly bearers of a leap into the heart of the matter. One thing that is discussed more explicitly and at length and in more varied dimensions in *Contributions* than anywhere else is the larger context of this opening to a radical transformation of thinking and being.[4]

The context for this transformative opening is our situation in the history of Western thinking. However, the word "history" here could mislead us if we are not aware of the distinction Heidegger makes between historiography (*Historie*) as the historical examination (*historische Betrachtung*) of a datable sequence of past events and historical mindfulness (*geschicht-liche Besinnung*) of the future as what comes to us through the ongoing unfolding of what has taken place in an originary—that is, a dynamically formative—way ("history" as *Geschichte*). Historiography, in its examination of the past, tends to create the illusion of objective distance. But in Heidegger's reflection on the history of Western thinking not only is the whole notion of objectivity put into question (along with subjectivity) but it also becomes quite clear that who, what, and how we are unfolds within the dynamic of this history, which "we ourselves are" (GA 45: 188, my translation).

Here, the possibility of a transformative "other beginning" for thinking opens up in an encounter with the first beginning of what we usually call Western thinking, an encounter that for the first time genuinely retrieves the movement of thinking in that early beginning. Consider a brief sketch of this retrieval.[5] We inherit our philosophical idea of "being" from the Greeks, who found themselves in the midst of beings without knowing what these beings *are*, without a knowing awareness of the "is" that these beings "are." This not-knowing astonished them and moved them to deep wonder at the being of beings. Attuned by this astonished wonder (which is, says Heidegger, the grounding attuning of the first beginning), they pondered this guiding question: What is a being? What is this beingness of beings? *What is it that is common to all beings as beings?* (Notice this quest for what is common to all rather than what is unique in each; this will become rather important later in the discussion.) The thrust of this questioning is to conceive being from out of some aspect of an understanding of beings. Thus, from Anaximander through Plato and Aristotle, the response to the Greeks' question gradually emerges in the determination of being as the presence of beings. Taken as what is constantly present in common to all beings, it grounds beings in their being, their presence. As the history of Western philosophy unfolds, being is not only differentiated and set apart from beings conceptually but is also reified in its constant presence. If being is presence, what is *always* present is most (in) being; it *always is*. It thus becomes the being that can ground the being of all less constant beings, not only as what is common to all of them but also as The Being. Being is here first differentiated from beings, as their ground, but the differentiating move itself is not explicitly thought or questioned either by the Greeks or in the subsequent history of metaphysics, the centuries-long history that multiplies names and interpretations of being (WHD 136; WCT 224). Moreover, that philosophical differentiating of being and beings

(the creation of what in hindsight can be called the "ontological difference")
is forgotten. Subsequently, the grounding function of being thought as a
being, as *The* Being, is simply assumed. Its meaning does not become a
matter for further questioning. And the notion of an *origin* of (the idea
of) being remains unthought and—within metaphysical parameters—un-
thinkable. There cannot be an origin *of* the ground and presumed origin
of beings (especially after this mode of thinking is adopted and used by the
medieval monotheistic philosophers) (GA 45: 152–80, 205–6; GA 65: 75–77,
232–33, 423–24/CP 52–54, 164–65, 298–99; WHD 98/WCT 152).

Following Heidegger this far, we can see how the guiding question of the
Greeks led them along; this insight into the guiding question also brings
its "answer" into question. As our questioning-thinking unfolds, being (its
origin, its meaning, and especially its function as ground) can no longer
so simply be taken for granted. This guiding question of the early Greek
thinkers, when *explicitly* thought as such, evokes what Heidegger calls the
grounding question of an other beginning: what is the meaning and arising and holding-sway of being itself? This is called the *grounding* question because it inquires into the ground of being itself. But what could
that "be"? How can there be a ground of what has for over two millennia
been taken as the highest and ultimate ground? What begins as a rather
straightforward sketch of the historical origin of our received notion of
being takes here a rather startling turn. To think that far already suggests
that ground and grounding may not "be" what we have assumed they are.
And perhaps being, if not ground, is . . . *what?* If we must ask those questions about the presumed ground of beings, we must also ask, What about
beings themselves? We (philosophers especially) hardly even give them, as
such, in their own places, a thought.

This thoughtlessness about beings is no accident. It has deep roots in
our philosophical heritage. Even though the guiding question of the first
beginning was about beings, once their beingness was conceived and determined as their rising into presence and into view, each unique being
becomes less and less significant. Being, modeled after beings though it be,
rules. "[T]he more questioning the question becomes and the more it brings
itself before beings *as such* and thus inquires into beingness and is consolidated into the formula τί τὸ ὄν, the more τέχνη is in force as what
determines the direction. . . . In order for Plato to be able to interpret
beingness of beings as ιδέα, not only is the experience of the ὄν as φύσις
necessary, but also the unfolding of the question under the guiding thread
of . . . τέχνη" (GA 65: 190–91/CP 133–34). What goes on here? In astonished

wonder the Greeks came up against beings as beings, rising up (φύσις, *physis*), revealing themselves (ἀλήθεια, *aletheia*, truth as disclosure), coming forth into view (οὐσία as εἶδος, *ousia* as *eidos*, presence as the look or appearance). But the dynamic rising and disclosing gets consolidated in the question τί τὸ ὄν (*ti to on*), "what [is] a being?" and τέχνη (*technē*, making) takes over.

Technē, which originally means relating to beings so as to understand and preserve their being, decisively shapes the first beginning in such a way that astonished wonder at the beingness of beings gives way to a manifold change in both understanding and action. This emerging Greek understanding aims at being—at the constant presence common to all beings. After Aristotle this beingness is grasped as the union of *morphe* and *hyle*, form and material, adding another layer of meaning to what can be thought of as what is "in common" to beings; later this commonality emerges as *essentia*, essence. All later forms of metaphysics, each in its own way, play out these distinctions. Not only does this readily converge with the monotheistic idea of The Being as divine maker (the ultimate technician, in the literal sense of the Greek), but it lays the ground for many later developments that go way beyond the domains of philosophy and religion narrowly construed. Along the way, ideas (instead of *physis*) and representability (instead of *aletheia*) become the measures of knowledge and truth. Beings become objects of representation; truth becomes correct statements about beings; humans begin to think of themselves as rational animals; astonishment, wonder, and questioning give way (in philosophy and eventually in modern science) to a drive for calculable knowing, uniformity and certainty; *technē* (preserving-making) becomes technique, machination (GA 65: 93, 109, 191/CP 64–65, 76–77, 133–34). This begins to sound rather familiar, as well it should. It also becomes, perhaps for the first time, genuinely questionable. We asked, What is being's origin, what is its ground? In one sense its origin is the creative thinking that took place with the Greeks. But then is being "itself" historically contingent and *without ground?* Yes, if ground is understood as it usually is, metaphysically. "*That* beingness was grasped as constant presence *from long long ago* counts already as grounding to most people. . . . But the inceptual and early character of this interpretation of beings does not immediately mean a grounding. . . . [T]his interpretation is not grounded and is ungroundable—and rightly so, if by grounding we understand an explanation that goes back to another being(!)" (GA 65: 195/CP 136–37). What then? Was the original conceiving of the ontological difference sheer groundless invention?

Let's take the questioning deeper. Our taking "being" and its beings for granted rests on an earlier forgetting that Heidegger calls *Seinsvergessenheit* (forgottenness of being), the forgetting of the originary move whereby being was first differentiated from beings. As soon as we mindfully consider the first beginning, however, that which was forgotten begins to emerge, but not in the way one might expect. The original conceiving of being is thought, but "being itself" cannot be found, no matter where we look or how carefully we think. This realization shifts us into an awareness of what Heidegger calls *Seinsverlassenheit* (abandonment of being). But if being serves as ground, then where is the ground now? No-where, apparently. No-thing as well. That is, "being" cannot be found as a being; in fact, being *is* not, at all. But if that is the case, if we can see that the *ontological difference has no actual ontological import*, then the function of being—in whatever guise—can no longer be taken for granted. The security and certainty of ground and grounding—the traditional function of being—is shaken not *by* us but *to* us, apparently (so it seems at first) by "being itself." That is, things *are*, are they not? So somehow, we think, "being" must *be!* But this no longer seems so sure, so well grounded. But wait a minute, that's rather slippery, too. The ultimate ground should not need further grounding. What "is" *being*, that it can now refuse to manifest so handily as ground? Did I not just say that being is not a being? At this point it must be said that this is no longer the guiding question of the first beginning but an *emerging* questioning of ground, which Heidegger calls the *grounding* question, that opens up within and toward an other beginning for thinking. It calls on us to confront abandonment of being and think what was *unthought* in the first beginning and subsequent history of metaphysics. It calls us to an encounter with grounding that is not a ground but rather *Ab-grund*, ab-ground, absence and staying-away of ground in a strong and dynamic sense. This is one way into the thinking that opens toward an other beginning in play with the first beginning of Western philosophy. Notice that the "other beginning" is not a consequence or subsequent result of the thinking of the first beginning. The possibility of an other beginning opens up *within* the thinking of the first beginning in its creative power *as beginning*. At the same time, the thinking of the first beginning emerges *as such* only under way within the opening of the possibility of an other beginning. Transformative possibility begins to move and emerge within careful being–historical thinking. And since we are that history, the way we deal with the contemporary world also comes within the scope of what calls for being–historical thinking.

The dominance of *techne* in shaping the first beginning effaces the uniqueness of beings in favor of what, held in common to all, can be represented by ideas. This technical dominance and the ensuing consequences in the history of metaphysics are no mere philosophical abstractions. If we had lived in medieval Europe, we would have assumed that beings are God's creations. God is being, *the* being grounding the being of all other beings as their maker, their creator. If we were well-educated Enlightenment-era Europeans, heavily influenced by Cartesian rationalism, beings would be thought of as extended substances to be known through clear and distinct ideas, grounded on the (presumably) undeniable existence of nonextended substances (mind and perhaps also God). All extended substances are in principle measurable and calculable, which means also controllable. Science, shaped not only by Descartes but also by Francis Bacon and others, comes to be understood as the means by which to firmly secure the god-given human domination and control of nature (GA 65: 111, 131–32/CP 77, 92).[6] And so it goes through the various permutations of metaphysics, with beings grounded on some idea of being, while that idea of being, just as it was in the first beginning, is determinable from the understanding of beings that is in play. So we ask, What are beings now in our technology-driven era? Heidegger opens this field of questioning most clearly in "The Question Concerning Technology," though he had begun to think along those lines already in *Contributions*.

The question is, What holds sway in the emerging and arising of modern technology as such? We are quite used to thinking of technology as our possession, as some range of available means to be used in attaining our ends. Reflecting on this instrumental means-and-ends definition of technology pulls Heidegger's thinking into a consideration of causality, specifically, of the four modes of causality outlined by Aristotle. What holds material, formal, final, and efficient cause together? They are ways of being responsible for bringing something forth into appearance. *Poiesis* (bringing-forth) may manifest as *physis* (self-arising) or *techne* (making); both of these are ways of revealing a being (GA 7: 9–13/QT 4–12). Here, a key insight emerges.

Technology, even modern technology, is a *way of revealing*, not merely a set of tools subject to our control and mastery. Modern technology, however, reveals things in a manner that is decisively different from other ways of revealing them. It is not at all a bringing forth that preserves something's own character, which was the Greek ideal, but rather a setting-upon nature that challenges all things to be constantly on hand for some predetermined use. Things are represented, ordered, and calculated in advance. They are interchangeable for any given use. They are disposable in at least two senses: disposing as setting in order and disposing as discarding the expendable. At the beginning of "The Question Concerning Technology" Heidegger urges us to pay particular attention to how language carries and reveals the matter for thinking. At this stage of the discussion he calls our attention to some of the ways that we describe the things around us. Almost fifty years later, what he says is startling in its accuracy. We can all too easily elaborate on it from our own experience and reading. Earth and soil: mineral deposits or land ripe for development. Farming: agribusiness. Food: nutriceuticals. The Rhine or the Mississippi: hydroelectric power source or mere source of coolant for a nuclear power plant. And even the river "itself" is only something to be viewed and photographed by a tour group organized and herded there by the travel industry. What emerges in this language is not something random or accidental. Notice that the river, the soil, the farm, and the food are subject to a particular way of being revealed. Caught in this web of interlocking processes that order them to stand by for disposing, beings are thus revealed as *standing-reserve*, a phrase that says, "The way in which everything presences that is wrought upon by the revealing that challenges. Whatever stands by in the sense of standing-reserve no longer stands over against us as object" (GA 7: 15–19/QT 14–18).

All things can, it seems, now be represented as even less than objects, as mere material for use and disposal. And this is just where the question of what holds sway in technology converges with the historical unfolding of the question of the meaning of being in the thinking of the first and other beginnings.

> The planning-calculating makes a being always more re-presentable, accessible in every possible explanatory respect, to such an extent that for their part these controllables come together and . . . in the moment when planning and calculation have become gigantic, a being in the whole begins to shrink. The "world" becomes smaller and smaller, not only in the quantitative but also in the metaphysical sense: a being as being, i.e., as an object, is in the end so dissolved into controllability that the being-character of a being disappears, as it were, and the abandonment of beings by being is completed. (GA 65: 494–95/CP 348)

So the thought of abandonment of being, arrived at earlier by way of coming to grips with the historical unfolding of the original idea of being, also emerges here in mindfulness of our ways of understanding, naming, and interacting with beings.

Being, the constant presence that grounds (our understanding of) beings, has had many names: forms, idea, substance, God, mind, noumenon, absolute spirit, and others. All of these say *ways of revealing* beings, whether as shadows of the forms, the unity of form and matter, creatures of God, clear and distinct ideas, extended substance, phenomena, and so forth. What about now, when beings have been reduced to less than objects, in standing-reserve? What now is *being?* Heidegger gives us the word *Ge-stell,* which says a gathered setting-upon and setting-in-place, or *enframing.* It says the ways in which we are corralled into a mode of being that challenges us to calculate, manipulate, and order all things into the interlocking webs of the standing-reserve. Beings have their standing only as enframed in this way (GA 7: 20–22/QT 19–21). The word Heidegger uses to say this in *Contributions* is *Machenschaft* (machination); this usage makes it a bit easier to see the connection to the history of the idea of being. When *technē* shaped the Greeks' answer to What is a being? it laid the ground for the ensuing history of metaphysics to culminate in *technē*-at-an-extreme, shorn of any sense of things arising in themselves (in *physis*), much less of wonder at the mystery of this arising. Instead, beings are now revealed through machination, which enframes everything as representable, calculable, orderable, and disposable, with no conceivable limit on the degree of quantification and control that can be expected. Science itself becomes an adjunct to machination, subordinated to the claims of ordering all things for production; it provides the specialized and ever-increasing refinement of rigorous accuracy that is called for. With no thought of conceivable limits, all questions are merely problems to be solved (GA 65: 108–9, 135, 145, 155/ CP 75–77, 94–95, 100–101, 107; GA 7: 20–24/QT 19–23; DT 45–46, 51).

At first it seems that we, who presumably benefit from all this planning and producing, are in charge of it. Having seen and dealt with the "human resources" (formerly "personnel") offices at our workplaces, we know that this is not necessarily the case. We, too, are subordinated to the compulsion of planning and production. Heidegger alludes to this with the example of "the forester . . . [who] is made subordinate to the orderability of cellulose, which for its part is challenged forth by the need for paper" for the newspapers and magazines that mold public opinion in precalculated directions (GA 7: 18–19/QT 18). "Downsizing" is a word that has recently entered the English language, referring to the disposal of expendable human resources. At an international conference on AIDS not long ago a prominent economist argued against the notion that it is not cost-effective to treat HIV in so-called Third World countries by telling his fellow scientists, "[AIDS] doesn't just kill workers, it kills young adults and young adults make children or raise children—*human capital. When you take that into the equation*, you find a very different impact on the economy."[7] We ourselves are indeed very close to being little more than units in a standing-reserve of human capital, apparently trapped in a situation in which any other way of thinking and of relating to beings seems highly unlikely. Why? Because of the way in which the compelling character of this one way of operating drives out even the hint of other options as either (1) not thought in the first place or (2) unreasonable (within this technical notion of what reason amounts to) or (3) impractical (i.e., not usable within the framework of calculable control) (GA 7: 26–29/QT 25–28; DT 56).

It is at this point in "The Question Concerning Technology" that Heidegger mentions a "saving power" that, from within enframing, could emerge and reopen other, incalculable possibilities. That so-called saving power is nothing reaching in from elsewhere to transform the situation. It converges with the thinking of the first and other beginning in *Contributions* as another way to name an opening toward transformative possibility. As opened up in the shorter essay on the nature of technology, this possibility (of an other beginning, of something radically different from entrapment in enframing) emerges from within a fundamental ambiguity in being's holding-sway now as enframing machination. This ambiguity has already been hinted at, in that machination is a way of revealing beings, a way of revealing that, however, follows upon forgottenness of being in such a way that it manifests abandonment of being in our thoughts and other acts. The dominance of *technē* in shaping the Greek notion of being is raised to exclusive domination, resulting in "a human epoch in which 'technicity'—the *priority* of the machinational, of the rules for measuring and of procedure . . . necessarily assumes mastery. The self-evident character of being and truth as certainty is now without limits. Thus be-ing's ability to be *forgotten* becomes the principle, and the forgetfulness of being that commences in the beginning spreads out and overshadows all human comportment" (GA 65: 336–37/CP 236). As the forgetting and abandonment of being first now becomes thinkable, a deeper forgetting also emerges, a forgetting of something so long hidden as to be not just forgotten but rather utterly unthought.

Lurking behind the forgottenness of the positing of the difference between being and beings is the deep hiddenness of the *arising* of beings (and thus also of being, which, from the beginning, is the being *of* beings). In their wonder at the beingness of beings, the Greeks did not inquire any further into the coming-to-be of beings or, in the language of *Contributions,* of be-ing, of the enowning that holds sway in all appearing, bringing each being into its own in the dynamic play of arising. In hindsight we can almost imagine that the incipient thought of *physis* and *aletheia* could, perhaps, have opened this matter to questioning, but that was not the path followed by the thinking of the Greeks or taken by the history of metaphysics after them. Thus, over two millennia later we have Heidegger, in *Being and Time,* telling us that in spite of our average everyday understanding of the linguistically pervasive "is" and our functional understanding of present-at-hand beings we not only do not understand the meaning of being, we do not even have the slightest inkling that there is a question to be posed about it or about its origins. What about this "average everyday understanding," when thought from within the thinking of the first and other beginning?

I am not asking here how that understanding is explained in *Being and Time* (that comes later in this chapter) but about how the "average Joe" thinks of beings, day to day. What comes to light must, of course, be uncovered from behind the vagueness and distraction that, for most people, keep this kind of question from ever coming up. However, a little thought reveals that the history of being—Western metaphysics as discussed by Heidegger—holds sway right out there on Main Street, on Wall Street, and out in the back yard with the Bud Light and the gas grill. The pervasive and generally unquestioned assumptions about "the way things are" are a confused, unarticulated mélange of the current dominance of enframing (techno-calculative thinking) entwined with older concepts: God and his creations, mind and body, subject and object, ideas and substance. Ask the man a few houses down the street how he can bear to see the misery of his dog, chained for years to her doghouse, and you may hear, "Dogs don't have feelings like we do." (Cartesian dualism rears its ugliest head.) If he is pressed harder or feels defensive, you may hear, "That's the way God made them." Ask the corporate CEO how he can justify a downsizing decision that cuts several thousand jobs while perhaps tripling his own take-home pay, and he may well (if he is not hiding behind the fifth amendment) say, "The shareholders are best served by my making this difficult decision, which will allow us to increase productivity per capita and per unit of cost." (Techno-calculative enframing obviously rules here, with workers grasped only as calculable cost-units, human capital, to be disposed *to* the planned objective or disposed *of.*) I could multiply examples, but this should suffice, along with a little thought. Many of the earlier determinations of the being of beings are still in play under the umbrella of enframing. Enframing allows the tendency toward objectification that arose in earlier conceptions to reach an extreme. All of this shows more concretely what I said earlier, that the history of being is not some datable sequence of events so much as it is a dynamically unfolding context that, with its variations on the idea of being as ground, shapes us or, as Heidegger puts it, that "we ourselves are" (GA 45: 188, my translation).

In the culmination of this history, when beings lose their standing even as objects and many of us begin to wonder whether we have any standing at all, the unease does not show up only in philosophy. We see in popular culture, too, the strife between a sense of groundlessness and a reactive compulsion to cling to a ground, to some ground, to *any* ground (God, morality, or even just profit). And in that strife there is a deeper malaise that is the strife between the sense (given in earlier metaphysical frameworks) of human meaningfulness or specialness or superiority and the growing suspicion or realization that we don't *count.* The result is a great deal of confusion and desperation that is not at all conducive to thinking. The current average everyday "understanding" of being stands in the way of posing even a preparatory question of the meaning of being (as in *Being and Time*), much less inquiring into what was left unthought in the first beginning or leaping into the open possibility of an other beginning.

Nevertheless, prodded along by Heidegger, it is now we who ask, What about the origin, the wherefrom and wherein, of coming-to-be? As the grounding question of an other beginning emerges from the attempt to come to grips with the first beginning, an attempt to bring this unthought wherefrom-and-wherein to language first becomes possible. This is challenging in that, as Heidegger affirms again and again, the truth of be-ing cannot be said directly in metaphysical (reifying, conceptual, systematic) language, and yet some merely invented or artificial language would also not *say* (show) the emerging thought. Instead, what is called for is a transformation of language and thinking that first opens up to and with the emerging thought of be-ing (GA 65: 78/CP 54). This will be a major topic of chapter 2. Here I want to give a preliminary sketch, just enough to indicate what Heidegger means when he says that *Being and Time* is thinking in the crossing. Thinking in the crossing to where, or to what?

In section 50 of *Contributions,* on the same page where Heidegger tells us that our experience of abandonment of being arises from machination's hold, he also says that this abandonment carries within itself an echoing hint of be-ing (GA 65: 107/CP 75). This opens up a distinction between the being (*Sein*) of metaphysics and its wherefrom-and-wherein, be-ing (*Seyn*). Be-ing, here, is not thought—as was being—from an idea of beings; that is, be-ing is not a being or a property of beings. Be-ing, he says, is "nothing at all, but holds sway" (GA 65: 255, 13, 235–36/CP 180, 10, 166–67). Be-ing's holding-sway is, in Heidegger's German coinage, *Wesung*, which says emerging as such, holding-sway "itself" (i.e., it does not mean "something holding sway" but holding-sway *as such*), or enduring coming-to-pass. Coming-to-pass also hints at *Ereignis*, another word used frequently by Heidegger that in this context says the enowning of what arises or comes to pass; this enowning again "is" not but *does*: "Enowning *enowns*," says Heidegger and is reducible to no being or event (GA 65: 349/CP 244). This proliferation of names—and these are not the only ones—for be-ing, for the unthought, says, each in its own way, the same, moving away from constant presence as ground into ab-ground, moving-thinking to engage with the retreat and staying away of ground. Here we encounter something elusive, something ungraspable in conceptual terms, no-thing that is not just nothing, something that calls for thinking that can say (i.e., show) be-ing while surrendering any claim to immediate comprehensibility (GA 65: 4, 14, 56, 64–65/CP 4, 11, 39, 45).

We must be careful not to mislead ourselves while we try to hear what these words say. Though they are not thought from out of some characteristic of beings in the manner of the first beginning, neither do these words name something extra, something *beyond* beings. Be-ing is not some sort of dynamic hyperbeing. Our forms of language, especially philosophical language, come to us from twenty-five hundred years of metaphysics. We must be mindful of the necessity and difficulty of avoiding reification, which would nullify the movement suggested in these *tentative* words. They evoke something that retreats as it comes forward, something that eludes our thinking grasp as it opens and makes a way for that thinking. They are spoken from a reservedness that accords with that elusive disclosure.

Just as the first beginning has its grounding attuning—astonished wonder at the being of beings—so too the other beginning has its own attuning. However, this attuning cannot be so simply named, as it is attuned by and to "something" that refuses to be represented in a name, as if it were a being. There are many evocative names but no grasping concept or representation either for "what attunes" or for the attuning. Some of these names are startled dismay (at abandonment of being), awe (in the face of the first hints of be-ing), and reservedness, but "there is no word for the onefold of these attunings" (GA 65: 14/CP 11).

Not forgetting that this is only a preliminary exploration, I ask, What is the relationship of be-ing and ab-ground in the thinking of the first and other beginning? "[W]hen being abandons beings, be-ing *hides itself* in the manifestness of beings. And be-ing itself is essentially determined as this self-withdrawing hiding. Be-ing already abandons beings in that αλήθεια becomes the basic self-withholding character of beings and thus prepares for the determination of beingness as ιδέα" (GA 65: 112/CP 78). When *aletheia*—truth—functions in the first beginning as unconcealing, as revealing and disclosure, it opens up a way of access to beings in their revealing themselves to perception and thought. Beings arise and emerge into the open to be examined in terms of *what* they are. However, this unconcealing is also fundamentally concealing, in that it opens up access to beings by closing off access to (the possibility of thinking) *arising and disclosing itself.* And this is left unthought altogether: the *originary concealing* that is always in play *in* the unconcealing, the self-withdrawing sheltering *of* the clearing and opening for the revealing of beings. This unthought matter of the first beginning harbors what now begins to be sayable, emerging into language, as be-ing and enowning. These names hint at and evoke the thought of "what" they name, but they resist any conceptual grasping that attempts to define and systematize them.

If we can respect the resistance and heed the resonance in the movement of thinking, we are shifted into an opening that clears the way for a transformation in thinking. From where we are now we can think abandonment of being (and be-ing) from under way on two converging paths: (1) the thinking of the history of being and (2) the mindful consideration of our epoch of enframing machination that is the culminating shape of that history. This abandonment reveals itself in the self-certainty that rejects ambiguity, denies all distress, refuses any limit or indeed any "no" or "not" in what is encountered, and covers machination with a veneer of "values." As abandonment of being reveals itself it also conceals itself in our enchantment with the apparent scientific and technological progress yielded by calculation and the rapidly accelerating movement from problem to solution to "the next thing," cutting off any questioning or doubt or hesitation (GA 65: 59, 117–23/CP 41–42, 82–86). This stupefying enchantment by technicity effaces beings in their unrepeatable uniqueness, "in the most ordinary publicness of beings that have become all the same" (GA 65: 238/CP 168). They are all the same, hence the continuous and desperate hunt for new experiences (fodder for the entertainment industry). We, too, are all the same in our orderability and expendability. Mystery and wonder seem to be gone, banished from consideration. But it is the nascent awareness of this banishing that moves us, even compels us, to engage more deeply with this mystery, with the unthought that is only now emerging into the possibility of an other grounding, an ab-ground.

Ab-ground names the movement of thinking into opening as it encounters what Heidegger calls "hesitant refusal of ground." *Hesitant* refusal indicates that this is not the flat denial of any thought of or inquiry into grounding. In fact, its movement is just the contrary of such closure. "Ab-ground is thus the in-itself temporalizing-spatializing counter-resonating site for the moment of the 'between,' as which Da-sein must be grounded" (GA 65: 387, 299, 311, 321–22, 342–43/CP 271, 218–19, 226, 240). This denial of ground, says Heidegger, opens up the possibility of our being -shifted from Da-*sein* to *Da*-sein, that is, from the *being* that is there in the midst of beings to being *t/here*. T/here, where? To be t/here in and, more aptly put, *as opening* for be-ing, making way for beings in their showing-forth, which also makes way for language and thinking. Thought carefully, ab-ground is none other than be-ing. (This will be taken up in much more detail in later chapters of this book.) To even begin to move into this manifold opening, this t/here, this "between," is to already undergo transformation. We cannot calculate or plan the transformations that may unfold but can only let ourselves be attuned as we thoughtfully attend to the paths that open up.

The language of Dasein here reminds us that it was in *Being and Time* that Heidegger first opened the way to this questioning of being and grounding. Heidegger is quite clear in what he says in *Contributions* that the attempt to open the question of the meaning of being in *Being and Time* is thinking in the crossing of the first and other beginning. What does that mean? We know that the thinking of the first beginning and the thinking of the other beginning are not separable as two distinct events, such that one could move *from* the one *to* the other. This historical thinking of the first beginning resonates within and enables an other beginning; historical thinking of the first beginning only takes place under way toward and within another beginning. To begin to think the history of being is to begin to think be-ing, to begin to hear the echoes of be-ing in the very language— itself historical—of being(s). To better understand the resonance of the first and other beginning, where be-ing is heard echoing (hinting) in the language of being, we need to reexamine *Being and Time* to see how it opens the region where this echo begins to be heard.

Being and Time: ATTEMPTING THE QUESTION OF BEING

The first step in the creative overcoming of metaphysics had to be taken in the direction by which thinking's posture is retained in one respect but in another respect and at the same time is basically led beyond itself.

> Retaining means: inquiring into the *being* of *beings*. But the overcoming means: inquiring first into the *truth* of be-ing—into that which in metaphysics *never* became a question. . . .
>
> The twofold character in the crossing, that grasps metaphysics more originarily and thus at the same time overcomes it, is through and through the mark of "fundamental ontology," i.e., the mark of *Being and Time*. This title is chosen on the clear understanding of the task: no longer beings and beingness but rather being . . . no longer thinking in advance but rather be-ing. "Time" as the name for the truth of being. And all of this as a task, as "being underway"—not as "doctrine" and "dogmatics." (GA 65: 182–83/CP 128)

Being and Time attempts to reopen the issue of being as a question, to awaken the kind of thinking for which such a question is a genuine issue. It takes place in relation to the entire history of being (metaphysics), in which naming being and explaining beings holds the central position. The necessity of *Being and Time*'s reawakening the question of the meaning of being situates the work within the forgottenness of the initiating thought of being, wherein being is first differentiated from beings and then rejoined to them as the being that remains constantly present to ground all that comes to presence (beings). As the history of metaphysics develops, the original posing of that distinction recedes into oblivion in the face of the manifold representations of being as that fundamental presence that grounds beings. Those can be thought as the various answers to the guiding question of the first beginning. But to place a question concerning the *meaning* of being is to begin to think in a way that is no longer limited in exactly the same way by the bounds of the original guiding question, What is a being? In raising the truth (disclosure) of being to questionableness, thinking moves on paths that begin to raise questions concerning the granting of presencing and arising as such: not just the ground of beings but the origin of being itself, the question of grounding as such. As these questions arise they contend with the guiding question's power to determine thinking. The thinking of the first and other beginning is set into motion as the early guiding and emerging grounding questions are thus set into play. Though the ground question is not explicitly raised as such in *Being and Time*, it is the attempt to think the question of the meaning of being that initiates this setting-in-play. The question of the meaning of being emerges as a necessary transitional question, and *Being and Time* thus prepares a crossing to the explicit thinking of the first and other beginning. How does this take place?

Contributions articulates *Being and Time*'s transitional role and its unexpressed transitional standing within the grounding question in terms of *temporality*, helping us to think this by-now very familiar matter in another light. Dasein—that is, each of us—as the t/here for disclosure is open as the "enduring of what is past and the taking-in-advance of what is to come," that is, in the language of *Being and Time*, Dasein's thrownness and throwing-open of and toward possibilities. The temporality of Dasein, in itself resonating with the historicality of being, is not a sequence of past-present-future (all of which have been thought only in terms of presence, of a sequence of fleetingly present "nows"). It is instead thought as dynamically relational being-in-the-world, as timing-spacing without grounding by some eternal "presence." The opening and "grounding" of the t/here as temporality intimates the deep holding-sway of be-ing in the enowning of Dasein, in its open relational presencing (GA 65: 73–76/CP 51–53). This sketch of the transitional place of *Being and Time* as thinking in the crossing should become more clear through a brief examination of some of the most pertinent sections of that text.

Being and Time unfolds as the phenomenological examination of an exemplary being, the one who, having language, is able to ask questions concerning its own being-t/here (*Da-sein*). As the text moves toward radicalizing Dasein's understanding of its own being as being-in-the-world, it reaches this decisive moment: "*Dasein is its disclosedness*," that is, its very being is the being of its t/here (GA 2: 177/BT 171). What does that mean? The t/here takes place as clearing or opening for disclosing, which is structured by the ways in which Dasein is disclosed to itself as being-in-the-world; it is the entire context of disclosive relations. In sections 31–34 of *Being and Time* this disclosive structure is first articulated as *Befindlichkeit* (finding oneself attuned), *Verstehen* (understanding), and *Rede* (discursive disclosure). In sections 65–68 the temporal character of this disclosive structure is explained along with another temporal aspect: falling. Finding oneself attuned is the coming to light of Dasein's thrownness (the phenomenal manifestation of its always-already-occurring and ever-changing facticity). To say it more straightforwardly, Dasein is already in such and such a situation and with such and such a tone or mood. Here there is already a certain understanding, that is, a disclosure of Dasein's situated possibilities, that can *now* be grasped or let slip away. So understanding and finding oneself attuned are co-original or equiprimordial (characterizing the whole of Dasein from the beginning). In projective understanding Dasein anticipates itself as possibility and thus stands forth as coming-toward itself. "Falling" names the way Dasein makes (things) present in encountering the ready-to-hand beings of its everyday activities. The temporality of Dasein's disclosive structure is not sequential, as it would be in the usual time-concept: finding oneself attuned, *then* falling into being-concerned with the ready-to-hand, *then* projective understanding of possibilities. Rather, Dasein's temporality temporalizes as the integral timing-spacing of the t/here, with the entire disclosive structure (including discursive disclosure, which allows Dasein to articulate its understanding in language) arising co-originally.

This temporalizing of Dasein's being already shakes the power of the guiding question of the first beginning to exclusively determine the course of this thinking in terms of grounding presence. Temporality *is* not but temporalizes. Temporality is thus neither presence nor ground but the way in which the t/here is cleared for disclosing. The transitional character of this thinking comes more fully to light if we consider language's role in structuring the t/here and then examine in more detail how making-present takes place (drawing on sections 15–18 and 78–81).

Dasein's being-in-the-world as the t/here for disclosure situates it in a web of disclosive relations in which beings show themselves to Dasein as ready-to-hand for some contextually understood use. They are meaningful and understandable within the whole web of significations. The understanding of beings as what is present is an abstraction from the always referentially understood character of things as being ready-to-hand with some role or function in the relational context. To even notice something as *only* "present" takes the effort of suppressing its usual meaning along with the entire context that supplies it. Dasein's most basic understanding of beings is of their showing up in an already-meaningful context of relationships. The crucial thing to notice is that there is no pure presence to be found other than what can be abstracted (in thought, in language) from the context that supplies functional significance to something. Beings (as beings in Dasein's world) are nothing in themselves apart from the context that supplies their significance. Only when there is a break in the web of relationality—when something is broken, unidentifiable, or "just doesn't make sense" in the surroundings—does something like mere presence-at-hand emerge. The thing is "just there" without other significance. Even then, it soon regains relational significance as something ready-to-hand for repair or the trashcan or simply retreating into the unnoticed background. The phenomenological explanation of the *derived* character of presence-at-hand from the everyday ready-to-hand surely shakes the assumption of the grounding character of presence. Notice the implication. "Being" is presence (in the entire history of metaphysics), but "presence" is an abstraction derived from Dasein's temporality, that is, from the dynamic disclosure of Dasein's everyday being-in-the-world.

What about time? It is already somehow implicated in any discussion of presence (as the present is ordinarily understood as a mode of time). At the beginning of *Being and Time* Heidegger says that the provisional aim of the book is "the interpretation of *time* as the possible horizon for any understanding whatever of being" (GA 2: 1/BT 1). Other than a few references to public time and Aristotle's notion of time, Heidegger saves most of what he says about "time" for the last hundred pages or so of *Being and Time*. The temporality of Dasein, however, is found discussed phenomenologically throughout the text. It would be fair to say that the book could well have been entitled *Being and Temporality* rather than *Being and Time*. I want to go a bit farther with the discussion of temporality in relation to time, to the point where, just when time itself is about to be discussed, the book breaks off.

Take this up in terms of what Heidegger says here: "The ecstatical unity of temporality . . . is the condition for the possibility that the there can be an entity which exists as its 'there' . . . 'open' for itself. . . . Only by this clearedness is any illuminating . . . any awareness, 'seeing' or having something, made possible. . . . Ecstatical temporality clears the 'there' primordially" (GA 2: 463–64/BT 401–2). What does that mean? First, the clearing is opening for Dasein's disclosure as being-in-the-world, structured by finding oneself attuned, understanding, falling, and discursive disclosure. The "discovery" of being-attuned is a finding out of our thrownness into a situation that is, in large measure, *already there*. Understanding is thought as *projective*, casting forward to what might happen or what we could do. Falling into the midst of beings, we make them present to us in their readiness-to-hand. Discursive disclosure opens the way for Dasein to articulate all of this, playing among all the other constituents of the t/here (e.g., in the tenses of language). So we could look at this one way and say, "There you have it: past, present, and future, all tied neatly together by language and thought." But it is that odd word, "ecstatical," that prevents our over-simplifying the matter in that way. Ec-stasis: standing outside. In this case, in the phrase "temporal ecstases" Heidegger is saying that each of these temporal constituents of Dasein's being-t/here "stands outside" itself. The constituents of Dasein's temporality are dynamically and inextricably inter-twined, with no fixed boundaries. "The future is *not* later than having been, and having-been is not earlier than the present. Temporality tem-poralizes itself in a future which makes present in the process of having been" (GA 2: 463/BT 401). Nothing in the structure of this temporalizing can be reified, that is, selected out and examined as if it were a being. Nor can temporality itself be reified. "Temporality 'is' not an entity at all. It is not, but *temporalizes*" (GA 2: 434–35/BT 376–77). It happens all at once, as dynamic relationality.

If we are honest, many of us will admit that when we first encountered this it sounded fairly weird. Why does it seem so strange? It simply doesn't fit neatly (or at all) with how we usually think of time, which is the com-monly shared understanding that Heidegger calls, variously, public time, clock time, parametric time, or derived time. (That last one, derived time, is an especially important clue.) What characterizes this understanding of time philosophically (and then shapes our ordinary everyday understand-ing as well) is that "it is a pure sequence of 'nows,' without beginning and without end, in which the ecstatical character of primordial time has been leveled off" (GA 2: 435/BT 377). But why do we presuppose that that is what time is? Obviously, it is not the only way to think of time; it is, however, the one that dominates our thinking, especially in the familiar guise of clock time or public time, which Heidegger also calls in *Being and Time* "they-time." Our thinking is not the only thing shaped by this notion of time; public time dominates nearly everything we do, as we race around from one appointment to another, meeting deadlines, finishing assignments "on time," filling up the time allotted "ahead of time" to various tasks. I look at my watch and say, "Now I have to do this or that," regulating my-self and my actions to accord with public, measured time. For this to work the measure has to be according to a fixed standard; if everyone's clock ticked at different speeds, presumably, the world as we know it would cease functioning. Of course, the precision of the current standard (the atomic clock at Colorado Springs) isn't necessary; earlier people looked at the movements in the sky and measured time, and they regulated their actions according to these movements, too. Measured regularity has just become more and more rigid and precise.

Is time, understood as measurable regularities reduced to a series of now-points, just the way it is? Is this somehow necessary because it simply reflects reality? No, the notion of time as a uniform, measurable sequence of nows has a historical beginning, emerging from within the unfolding of metaphysics, specifically, with Aristotle, whose account of time as what "is counted in the movement encountered within the horizon of the earlier and later" dominates all subsequent theorizing about time (GA 2: 35/BT 48–49). What, precisely, is it that gets counted? The nows, the fleeting moments of presence. "Thus for the ordinary understanding of time, time shows itself as a sequence of 'nows' which are constantly 'present-at-hand,' simultaneously passing away and coming along. Time is understood as akin to a flowing, one-way stream of these 'nows'" (GA 2: 556–60/BT 473–75; see also GA 65: 223/CP 156). This all sounds very ordinary and familiar. But if we think about it just a little bit, it is every bit as strange as *Being and Time*'s account of the temporality of Dasein.

What is this "now" that comes along and passes away? Try to find one, just one. Already it is gone. Try, in thought, to split it up; no matter how you imagine doing that, what is left is still just now. It seems a bit like a mathematical point (no extension in and of itself), but the now cannot be located on a grid like a point. The now "was," but it was not "somewhere." *Right now*, we can say the now is *present*, but if it is gone before we even finish reading this sentence, what kind of presence is that? And if, as metaphysics has it, being is presence, then what is the being of time? Non-being? But if that is so, why is time so powerful? Time, in our cultural and philosophical tradition, seems to stand outside the flux of events, beyond the coming, going, and changing of beings, serving as one of the two means by which we *measure* them (with space as the other). Everything has its stretch, its span of time and extension in space. But if time and space stand outside in that way, then one would tend to assume that they must somehow exist. But how does that make sense? Even the "space as empty container" metaphor is a bit difficult to accept (I will take that issue up later, in chapter 3), but the "flowing river of nows" is even more implausible. If the nows are flowing, they must be moving, coming and going. Coming from where and going to where? Either there is "something else" beyond time (in which case "time" originates in or is derived from that), or we must say of time, as of being, time is the time of beings. Or perhaps both (that discussion must wait for chapter 3 as well).

In *Being and Time* Heidegger sketches how this common understanding of time as an infinite series of nows is derived another way, from the "flattening" out of Dasein's dynamic, ecstatical temporality, which always unfolds in the context of relational, meaningful significations pertaining to the beings with which Dasein is concerned. The usual notion of time covers up all these relations, and the dynamic structure of temporality gets "*leveled off.* The 'now' gets shorn of these relations . . . and . . . they simply range themselves up along after one another so as to make up the succession" (GA 2: 557–58/BT 474). So time as it is ordinarily understood and as it shapes our everyday actions is a derived abstraction rather than an ultimately existing framework.

As has been often noted, the ending of *Being and Time* (which comes soon after the discussion of the derived character of public time) does not mark the conclusion of its stated project: to formulate the question of being so that it can become possible to answer it and to articulate its horizon as *time.* The final division of the book, which would have been about time, is missing, and, according to the brief outline given, it would have accounted for more than half the work. This is so in spite of the fact that *Being and Time* contains many things that could seem to be an *answer* to a question concerning the meaning of being. Of "being" we learn these things:

1. Historically and in linguistic usage being means "presence."
2. Phenomenologically, for Dasein, being is the readiness-to-hand and presence-at hand of the beings in Dasein's world.
3. Being also means Dasein's being as temporality, in the world, as the t/here for disclosure.

So on the surface it would seem that *Being and Time* fulfills its aim of formulating the question, What is the *meaning* of being, and what is the horizon within which this question must be thought? And it seems to answer the question as well.

But if the question has indeed been formulated and answered, why does *Being and Time* end the way it does, and why, after that, does Heidegger's thinking not go directly to the task of writing the missing division, as it was outlined, instead of spreading out into the many productive directions it took? After *Being and Time* Heidegger lectured and wrote at length on the history of philosophy as the history of being (including lengthy works on the great thinkers of Western philosophy), language, the nature of thinking, how we are in the world today under the dominance of technical ways of thinking and acting, and the nature of language. In the course of this extended work of thinking, speaking, and writing he included some thoughts on time, though they were usually somewhat brief and, without the context that *Contributions* would have provided, cryptic. What is said in *Contributions* about time, however, is nothing like what is outlined in the proposed "Division Two" of *Being and Time.* So I return to the question posed above: What prevents the ending of *Being and Time* from being a fulfillment of its project? This closely related question may also be asked: What prevents the proposed "Division Two" from ever being written as outlined and thus enacting the fulfillment of the stated aims of *Being and Time?*

One reason unfolds from the somewhat limited character of the beings that are the main focus of *Being and Time.* The phenomenological analysis basically covers two kinds of beings: Dasein and all the other beings in its world. Logically, that would seem to cover all beings. However, in the course of the book, beings not of Dasein's kind prove to be beings ready-to-hand and beings present-at-hand in Dasein's *world*, that is, in the web of significations by which Dasein understands and according to which it moves through its world (which itself is defined as the total of these significations). Dasein, as the t/here for being, is the opening for the revealing (self-showing) of beings. But what "beings" (seems to) mean here is beings that are such *only for* Dasein and that arise and appear as such. What about those that never show up, so to speak? What about, for example, the denizens of the deep wilderness? It seems dubious to simply define these "out of being." It would seem at the very least that—once again!—the very *meaning* of being that is in question and at stake here has been to a certain extent predetermined. Yet it is quite clear that to *question* the meaning of being will require clarity about any presuppositions concerning the matter. So, as Heidegger himself says at the end of *Being and Time,* the question itself has not been adequately or fully formulated, much less answered, and "the conflict as to the interpretation of being cannot be allayed, *because it has not yet been enkindled*" (GA 2: 576/BT 487). To leap directly from the phenomenological analysis of the beings of Dasein's world (all of them *at hand* in one way or another) to the proposed division on *time* could not yet be attempted.

That said, what kinds of beings are missing from any significant consideration in *Being and Time?* Earth, animals, and plants. Even in terms of the focus of *Being and Time* itself, there is something a bit odd here. However, this oddness is not so much a flaw in *Being and Time* as it is a reflection of its character as thinking in the crossing, as the transitional opening to the thinking of the first and other beginning, preparing the leap to the thinking of be-ing. As such, the thinking in *Being and Time* unavoidably mirrors and echoes the thinking of the first beginning and also unavoidably reflects some of its limitations. In that mirroring-echoing, however, we find ourselves already engaged, along with Heidegger, in opening up and thinking more deeply into our inherited notions about being, about ourselves, about beings, and about time. The first beginning, attempting to grasp being as the arising into presence and appearing of beings in terms of what is conceivable by us, could not *bear* the explicit thought of anything in the domain of concealing, withdrawing, or hiding except insofar as it could be dismissed as dissembling or untruth. So beings that aren't beings "for us" as well as earth—which resists being fully revealed in the web of significations called "world"—tend to fall by the wayside both in the history of Western philosophy (and religion) and even in *Being and Time*.[8]

Fairly early on it seems that Heidegger had a sense that he wasn't going to stay on the surface of the "world," that is, of what is captured in Dasein's web of significations. In later lectures and writings, from "The Origin of the Work of Art" (1935–36) on, there unfolds a long pondering on the "strife of world and Earth" that plays out the irresolvable tension between revealing and concealing.

There are many things in our lives that offer resistance to our conceptual grasping and controlling actions. Our own embodied existence is one of those things. Already in *Being and Time* Dasein's mortality is a major focus of discussion, couched there as a matter of meaning within Dasein's world. We, of all the animals, *know* that we will die, making our very being an issue for us. We think, we know, we question. But our concern and our questioning about death and about meaning have much to do with our having/being animal bodies and thus not just being of world but also of earth. *We embody the strife of world and earth.*

The temporality of Dasein as it is articulated in *Being and Time* (in the world, in a web of significations of things ready-to-hand and present-at-hand, and with other beings of Dasein's own kind) *is not yet enough* to open up time as the horizon of any possible answer to the question of the meaning of being. And this is not only because of the need to think and question regarding earth and earthy beings. Consider the form of the question: it is the question of the *meaning* of *being*. *Meaning* indicates that, unless we would simply adopt all prior assumptions (something Heidegger makes us cautious about from the start), the very notion of meaning will have to come in for close scrutiny. But that requires looking very carefully at how language and thinking work. In *Being and Time* Heidegger makes a fair start on opening up the matter of language (especially in section 34).

But *thinking* is placed less directly into question or opened up as an issue for thought. In 1927 Heidegger calls upon the methodology of phenomenology (though he is already thinking more deeply about what that means than did Husserl) with additional inspiration from theological hermeneutics (again, with caution and reservation). He is *beginning* to move beyond those philosophical-interpretive methods but has not yet begun to articulate *why* that is necessary. He has not yet called the very notion of "method" into question, as he does later.

Another related set of issues arises from the fact that these questions are emerging in the context of the attempt to think in the crossing of the first and other beginning of Western philosophy. I have already pointed out that the kind of language that articulates ideas about being emerged from certain assumptions about beings. While that is quite correct, it is only part of the story, and what is missing is crucial. David Abram, in his remarkable book *The Spell of the Sensuous*, makes a convincing case that a shift in our relationship with language was taking place *already as* the Greeks were first beginning to think metaphysically. He notes in particular the ramifications of the gradual shift from the prephilosophical world of the Greeks, where knowledge was passed on orally, to the acceptance and use of alphabetic writing. Within strictly oral cultures meaning is not something originating only in human speech and thought, but it is understood to arise within the larger animate matrix that we now refer to as nature: plants, animals, rivers, forests, stones. Thinking of this in terms of contemporary evolutionary understanding, our bodies and *all of their capabilities*, including thought, have formed "in delicate reciprocity with the manifold textures, sounds and shapes of an animate earth."[9] All that we can conceivably call "human" arose only in complex intertwining with the dynamic nonhuman natural world. Thus, Abram first undermines the notion that language is an exclusively human artifact or property. At the same time, we are brought to wonder what we have lost, once language and thought are narrowed to our exclusive possession.[10]

Abram also tells a story that—once we bring it into the context of the first and other beginning—helps us to catch a first glimpse of the complexity of the relationship between ideas about beings and the language that expresses them. I have already mentioned one of the pieces of this puzzle; it was the Greeks' fascination with *technē* that gave its particular cast to the developing metaphysics that first reifies "being," differentiating it from the "beings" that it then can serve to ground (in our thinking about those beings). Abram calls our attention not to *technē* in general but to a particular technology: alphabetic writing. We have been literate for so long that we have long since forgotten that writing is a technology, that it is both a *means of production* (the commonsense idea of technology) and a *way of revealing* (the deeper sense of the word that Heidegger calls to our attention). And *alphabetic* writing, in contrast to earlier pictographic writing, is a technology that, once it was accepted and used, fostered and enabled metaphysical thinking. Pictographs depend for their meaning on an explicit reference to something in the world, something they picture. The alphabets used in the West arose first from early pictographic writing in ancient Hebrew culture.[11] When this technology was brought to Greece, the sounds lost that tie and now were only just that: markers for humanly produced sounds. It was this change that gradually also shifted our attention away from nature, the sensuous and sensible world, toward the human intellect as the locus and presumed originator of meaning. We see those words "sensible" and intellect, "intelligible," and they call to mind the distinction that Plato worked so hard to inculcate by way of the Socratic dialogues.

Abram points out that it was the technology of alphabetic writing that enabled Plato to accomplish what he did and in so doing to shape our thinking from that time on. What this writing technology allowed was for thoughts and ideas to be set down in such a way that they existed in a form previously unimagined. They were relatively permanent and could thus be taken up at will and examined by anyone who could read, apart from the one who wrote them. Of course, the writer could do the same. "A new power of reflexivity was thus coming into existence, borne by the relation between the scribe and the scripted text."[12] This cleared the way for Plato to be able to carry forward with two closely linked notions: (1) that some of the "ideas" thus grasped and fixed actually had being of their own as the eternal forms and (2) that the human mind (or *psyche*) itself could—and indeed *should*—think and know the true and the real, that is, those *ideas*, apart from the messy, changing, often-confused body. So it is that the creation of the ontological difference (as I have sketched it, following Heidegger) *arises together with* very particular assumptions about ourselves and about language, meaning, and beings. Abram puts this very well, saying that "the new relation that Plato wrote of, between the immortal *psyche* and the transcendent realm of eternal 'ideas' was itself dependent on the new affinity between the literate intellect and the visible letters (and words) of the alphabet . . . accompanied by a concomitant internalization of human awareness . . . [that can] interact with itself in isolation from other persons and the surrounding, animate earth."[13]

Therefore, as we attempt to move forward with thinking in the crossing of the first and other beginning, it is important to realize that the question of the meaning of being is not a question that can be examined in pristine isolation from the powerful assumptions that arose along with that first grasping and reification of "being." The assumptions that have already come to light include (1) dualistic notions about our nature and about our relationship to the earth, (2) presuppositions about the nature of language and our relation to it, and (3) assumptions about what constitutes good thinking. I will return to the first point later in this book. The second and third items come in for scrutiny in chapter 2.

2

Thinking

Engaging Language's Way-making

In "A Dialogue on Language" Heidegger said that "reflection *on language* and on being has determined my path of thinking from early on" (GA 12: 88/WL 7, emphasis mine). After the considerations of the first chapter this should come as no surprise. In the first place, the matter for thinking in *Being and Time* is the question of the *meaning* of being, which already hints at a central place for language in transitional thinking. *Rede,* the discursive disclosure that is at the heart of language, is fundamental to all questions of meaning (GA 2: 214/BT 204). Further, when *Being and Time* is resituated in terms of the historical thinking of being in *Contributions* and the questioning concerning being goes deeper, the way that language carries out this thinking (and all thinking) comes into sharper focus as an integral aspect of what is in question. In *Being and Time* as thinking in the crossing the grip of the limiting power of the first beginning has been shaken. In section 68 of *Being and Time,* for instance, there is already an indication that since being (as the being of Dasein and of the being of the beings of Dasein's world) is relational and temporal, it thus "is" nothing in itself. The insight is that being becomes meaningful to us *only in language.* The question of the meaning of being (and of time as its horizon) is therefore not only a question concerning being but also a question of language, that is to ask, how does discursive disclosure, emerging as language, take place in such a way that "being" and "time," however they may be understood, come to have meaning for us? What is the nature of this creative energy of language?

Near the beginning of section 34 of *Being and Time* Heidegger says that discursive disclosure is co-original with understanding and finding one-self attuned, the other key elements of Dasein's temporality (GA 2: 213/BT 203). But earlier, the first time the three structural elements of the t/here were introduced, he had put the matter somewhat differently, saying that while understanding and finding oneself attuned are co-original with each other, they are both "characterized co-originally" by discursive disclosure (GA 2: 177/BT 171). This difference in emphasis is explained in further discussion of the temporality of the disclosive structure, especially section 68. Discursive disclosure is not temporalized in just one or the other of the temporal modes of having-been—either projecting-open toward possibilities or falling into concern with the ready-to-hand—but is what accounts for and gives an account of the significance and meaning of the entire structure, articulating this meaning most often—though not exclusively—in language. Dasein grasps its being in a world alongside beings as the web of significations gets articulated and thus becomes meaningful. The assumption of the grounding character of presence is shaken in yet another way here. Presence-at-hand is derived as an abstraction from the ready-to-hand. Likewise, familiar public, measurable time (the uniform linear sequence of nows) is derived from nonsequential, dynamically relational temporalizing (which again involves Dasein's interaction with and understanding of the ready-to-hand things in its everyday world). However, *the significations that bring Dasein to understand the ready-to-hand only become meaningful through discourse.* Further, to move from presence-at-hand to "presence as such" requires yet another move in language, by which temporal making-present is further abstracted out and reified in words. Presence as such is none other than *being as such.* Though *Being and Time* goes no further in its task of explaining time as the horizon of the question of the meaning of being, this is already a significant transformation in thought. Many years later, Heidegger says that carefully rereading *Being and Time*, especially section 34 (on language), shows clearly that it is already engaged in a transformation of thinking that (eventually) leaves metaphysics behind. Where it goes in that move is not yet indicated in *Being and Time* and is left nameless even much later (GA 12: 130/WL 42).

Both in the historical thinking that ponders the origin and various determinations of being in metaphysical thinking and in the phenomenological exposition in *Being and Time* we arrive at the same place: the realization that the being of beings—thought in terms of the ontological difference, as a being—is something that shows up in and only in language. That would also suggest that representing being as the ground of beings is what

produces that ground as ground. In terms of the thinking of the first beginning, thinking the difference of being and beings creates the very idea of
being as such. But at the time, that must have seemed much more like a
discovery. What Heidegger first names the ontological difference was not
itself *as such* (i.e., as a creative differentiating of one phenomenon into two
concepts), explicitly thought (much less questioned) in the first beginning.
Only the being of beings was put to the question; once the difference between being and beings was "discovered" (created in thought), the differentiating move that took place in the course of that thinking was deeply and
decisively forgotten, cast all-unknowing into oblivion. "The oblivion here
to be thought is the veiling of the difference as such. . . . The oblivion belongs
to the difference because the difference belongs to the oblivion. The oblivion does not happen to the difference only afterward in consequence of
the forgetfulness of human thinking" (ID 50–51, 116–17). Only thus could
"being" become what it did: constant presence assumed as ground for
over two millennia of Western philosophy and culture. How being is represented determines how beings can and will be represented; it says what
things are, what they can and will be, and, in so doing, implicitly says what
they will not and indeed cannot be (GA 12: 182/WL 88; GA 65: 83, 499/CP
58, 351).

    Bringing this within the region of thinking opened up in *Contributions*
sets the discussion in *Being and Time* in its place as transitional, preparatory thinking in the crossing to and within the thinking of the first and
other beginning. Reified being and parametric time arise from out of the
first beginning. As they come into question and as thinking attempts the
leap (in)to ab-ground ("wherein" there is nothing to reify and yet much
to be thought), it becomes necessary to try to find other ways of bringing
the thinking to language or, perhaps better said, to let language carry the
thinking. Thus in *Contributions*, from beginning to end, language itself is a
crucial matter for thinking. On its first page Heidegger says that the proper
title of the book, *From Enowning*, "is not saying that a report is being given
on or about enowning. Rather, the proper title indicates a thinking-saying
which is enowned by enowning and belongs to be-ing and to be-ing's
word" (GA 65: 3/CP 3). What this means will need to come forward gradually as the discussion proceeds, but one thing is clear already, namely, the
heavy emphasis on language. Emphasis is placed not just on be-ing but
on be-ing's *word*, on the language that carries and evokes the thought of be-
ing. The last section of *Contributions* is entitled "Language (Its Origin)" and
includes this thought: "Language is measure-setting in the most intimate

and widest sense . . . and . . . ground of Dasein" (GA 65: 510/CP 358–59).
So, on its first and last pages, *Contributions* emphasizes the central place
of language in the thinking of be-ing, the transformative thinking of the
first and other beginning. The transformative possibilities of this thinking
arise in the thinking-saying of be-ing, that is, in language. But since language, too, is undoubtedly historical, and our Western languages are thus
strongly shaped by the metaphysical notions of the first beginning, would
this not require a transformation of language itself? Yes, "a transformation
of language is needed that we can neither compel nor invent. This transformation does not result from the procurement of newly formed words
and phrases. It touches on our relation to language" (GA 12: 255–56/WL 135).
It does more than just touch on it; if the thinking goes well, it may well
put in play an experience with language that, says Heidegger, "overwhelms
and transforms us" (GA 12: 149/WL 57). This is not some other transformation than the transformation in thinking already discussed in chapter
1: we are digging deeper, and as we go the matter broadens. As I said there:
change one key thing, and everything changes.

Being shows itself in and only in language. Be-ing, too, shows itself in
language—whether it only manifests in language must for now remain an
open question. The attempt to hear the echo of be-ing in the language of
being(s) calls for making language and thinking explicit (and intricately
intertwined) matters for thinking. But what of this thinking attempt? Something very strange shows up here. How can thinking call into question the
way in which language arises and holds sway with such power when, as
far as we know, thinking itself is always embodied in some kind of language? How can thinking let itself be attuned to be-ing when the language
in which thinking takes place is the historically determined language of
beings? That is, the language is metaphysically expressive, linear, and dualistic (subject or object, presence or absence, being or nonbeing). How is
it that language has this duplicitous power to both reveal and conceal matters of decisive importance? "This transformation of language pushes forth
into domains that are still closed off to us because we do not know the truth
of be-ing" (GA 65: 78/CP 54). The truth of be-ing: be-ing's ownmost way of
disclosing "itself." If that seems impenetrable at this stage, it is because it
attempts to say something nonmetaphysical within an intrinsically metaphysical language structure. What is the "it" of this pronoun "itself" if it
"refers to" be-ing? Noun, pronoun, reference, subject-predicate—all these
things lean heavily toward encouraging this understanding: be-ing is something that somehow exists. But be-ing is not a being, and its "grounding"

is ab-ground. The strangeness of this language of be-ing and the difficulty we have in understanding it presses us to question language in yet another way. If be-ing is to be thinkable at all, it must somehow be sayable. How does or can that take place, the sayability and thinkability of be-ing? How can it happen in the face of the power of metaphysical language (language oriented to the presupposition that being is presence) to limit what is thinkable and sayable? It would be a fairly outrageous stretch of the imagination to think we could completely revise language to delete nouns, pronouns, and anything else that tends to encourage the assumption of the substantial self-existence of beings and being. That is quite obviously so far-fetched that we can safely say that it is not what Heidegger means by a transformation of language. And Heidegger says in various ways, at various times, that we are not looking for new words to invent or new linguistic structures. Even though that is not in the cards, the transformation he has in mind actually runs much deeper and broader. As was said above, it has to do with our relation to language. But our relation to language, shaping our thinking from top to bottom, mediates our relationship with *everything*. Here it is again: change one key thing, change everything. Specifically, here, change our relationship to and understanding of how language works and along with it the nature of its saying power (and also of what is being said and shown), and inevitably we will also change. But this raises another question. How can this happen in the face of metaphysics at its extreme, the grip of enframing that shapes not just our dealings with beings and with each other but also our language and thinking? We have a thick knot of questions here. All of them point toward the necessity of thinking more deeply *about thinking* itself, and that means also thinking more carefully about the language that opens the way and carries the thinking.

How do we enter this knot to attempt to begin to unravel it? I will take as a strong hint two comments made by Heidegger in the mid- to late 1950s. In one case, when asked why he had stopped referring to his thinking as either phenomenology or hermeneutics, he said, "That was done . . . in order to abandon my own path to namelessness" (GA 12: 114/WL 29). How can that be, we might wonder? Doesn't all philosophical thinking proceed according to some method or at least some identifiable plan? Apparently not. Where, in fact, does that question come from? It comes from our philosophical heritage, of course. (Where else, indeed?) And as it turns out, while that heritage gives us much, it also hinders *this* thinking, especially in its techno-calculative guise now at the extremity of the historical unfolding of metaphysics. Therefore, Heidegger also advises us that if we wish to *learn thinking*, we must radically unlearn the traditional methods of and presuppositions about thinking (WHD 12/WCT 8). But this "unlearning" is not going to happen by accident or whim or by some kind of willful choice, such as "I will start from scratch, with no presuppositions at all." How likely is it that such a free-floating act of will can succeed? "Unlearning" in itself implies that some *learning* has already taken place. In this case the learning goes so deep that it can fairly be said to be embodied in our very sense of who and what we are. From the time we begin to hear and learn language we are also learning how to use language not just to communicate with others but also to "talk to ourselves," to think. We absorb the presuppositions embedded in the grammar and syntax of our native tongue long before we are able to think about them, much less to call any of them into question. At some point, most reasonably well educated people encounter the idea that language enacts cultural predilections and presuppositions. Nowadays this most often comes up in discussions or debates about multiculturalism and cultural relativism. But what is at stake there is something significantly different from what Heidegger is putting into play here. He is saying that it requires effort to (1) learn our currently dominant mode of thinking well enough, and with enough clarity, so as to begin to see its power to limit what is thinkable and (2) unlearn that traditional thinking, which is not the same as abandoning or willfully rejecting it, but learning to let it be optional, which is more difficult than it might seem. Only then will we be able to learn another, radically different way of thinking.

LANGUAGE AND THINKING AT THE EXTREMITY OF METAPHYSICS

The way that the history of Western philosophy unfolds from the first beginning not only produces its many interpretations of how being functions to ground beings, it also develops ways, methods, and standards for the thinking and language that express those determinations of being and beings. Early on we have Plato's elevation of the intelligible over the sensible and Aristotle's hierarchy of the sciences, with mathematics and metaphysics ranking highest in precision and likelihood of accuracy. Later we have various decisive and influential developments in formal logic and epistemology, especially when we move through the medieval period, with its synthesis of monotheism and Greek thought, and then come down to the modern era, with Bacon, Descartes, Newton, and others. Even if we are not students of the history of Western philosophy, the results of this are

not by any means news to us. We all know, for the most part, at least generally, what the following words mean: method, reasoning, argumentation,
objectivity, evidence, criteria. We know that methods can vary, along with
the criteria for what counts as evidence and to some extent the rules or
standards for good reasoning. But the received assumption is that without these things we would be left with blind emoting, or "mere subjectivity," or, at best, substandard or specious reasoning. With these elements of
what makes up our idea of "good thinking," on the other hand, we can
hope to produce those things at which good reasoning aims: representations, concepts, definitions, facts, truth, and theories. All of this has about
it a sense of rightness about both the process and its results. It is familiar,
it is relatively accessible; it is, in fact, downright comforting. Why question any of this? Heidegger does so for at least two reasons. One is what I
have already discussed briefly just above: if metaphysical (reifying) language
is unlikely to be able to bring the thought of be-ing to language, then it
would seem likely that the kind of reasoning that developed in the ongoing process of refining that language can only continue to produce more
reifying language. That is, language and thinking are very hard—perhaps
even impossible—to separate even in principle, much less in practice. The
other reason Heidegger calls current thinking practices into question is the
dominant form that thinking has assumed most recently in enacting the
rule of being as enframing. Both of these points call for some clarification.
I examine the latter first and take up the problems with the methodology
and standards of what we usually take to be good reasoning or good thinking later in this chapter.

In a short presentation to his neighbors, published as "Memorial
Address," Heidegger very concisely and clearly sketches out the characteristics of the dominant mode of thinking. "Memorial Address" is so read-
able and accessible and in some ways so simple compared to most of what
Heidegger published that it is often used as introductory material for undergraduates (for which it works rather well) and overlooked by us the rest
of the time. That would be a mistake and a loss. This short lecture speaks
from within the heart of the transformative thinking of the first and other
beginning. It lays out the danger of the dominant way of thinking in our
age, and it opens a way out with two simple but incredibly profound hints
about a radically different way to think. First, consider the dominant mode
and its danger. Since what Heidegger says in "Memorial Address" speaks
from within the thinking of the first and other beginning, we can expect
that this danger will resonate with the danger already encountered in the

account of enframing in chapter 1. There we realized that we are on the brink of becoming no more than mere interchangeable units in a standing reserve of "human resources" or "human capital." How could that be? How could we let that happen? It grows out of the very sense of comforting and compelling "rightness" that reasoning and planning give us. Standards of reasoning have converged more and more tightly with the aims of control and production (from *technē* to technicity, with all the power of contemporary technology now in play). What seems to us to be the best—most productive—thinking has now become very specialized. It begins by taking given conditions into account to serve very specific purposes so that we can count on the definite results that we want. Heidegger calls this mode of thinking *calculative* thinking, because whether or not numbers or computers are involved, the process is akin to mathematical process. In ordinary, everyday language, too, we sometimes speak of someone's "calculating" ways; the person in question is unlikely to be a mathematician. We know quite well what the sense of the adjective is: being clever in a way that allows one to manipulate people and situations to advantage. Calculative thinking, as it is sketched by Heidegger, is characterized by speedy and efficient problem solving, with definite goals and methods of attaining them. That doesn't sound very alarming. On the contrary, it sounds quite normal. It is just business as usual. But there are wide-ranging ramifications. Again, this is not news, but Heidegger is asking us to *stop* for a minute and see this clearly as it is and what it means for us. To do so is already to be shifted ever so slightly outside the bounds of calculative thinking, which, says Heidegger, never stops to consider the meaning of its process or its products. Some of the ramifications have already come up in the discussion of the place of enframing in the historical thinking of being as the extreme, culminating, self-destroying (in bringing beings to lose their own standing as such) culmination of metaphysics. Beings become less even than objects, being taken as mere resources to be stockpiled and used as needed. Nature itself, says Heidegger, "becomes a gigantic gasoline station," and our own relation to the world becomes a strictly technical one. Already in 1955 Heidegger had taken note of some comments by scientists that pointed forward to what we now call genetic engineering, which produces genetically modified organisms. That this would develop to the point where vast tracts of the corn and soybeans grown in North America are planted in genetically modified organisms, with the result that we are probably all eating foods that contain them, may not have come as a surprise to him. The fact that the various components and modifications of living organisms can now be patented in order to control who may profit from these developments certainly stands as a vivid example of beings and even life itself having been unquestioningly taken as stock, as standing reserve (DT 46–53).[1] It is quite obvious that this has major implications for how we view ourselves and each other. But there is even more to be concerned about if we are interested in attempting to *think*.

Why, we might wonder, is there so little effectively expressed concern about this? Look around at the barrage of information that assails us very nearly continuously: radio, television, film, roadside signs, newspapers, magazines, the internet. All of this is captivating—even more so when it holds out the promise of solving our various individual and group problems. We can, immediately, at any time, find out how to improve our "lifestyle" or, alternatively, be entertained sufficiently that we no longer worry about it. Even contemporary psychology and quite a bit of popular spirituality are couched in calculative, problem-solving, efficiency-oriented terms. Beauty, sexual fulfillment, and enlightenment are all on offer in the marketplace: self-hypnosis tapes, weekend workshops on shamanic soul retrieval, a weight-loss plan for every taste, "win a free makeover!" All of this creates a surface aura of encouraging individuality while at the same time channeling us all into slightly different versions of the same track. One of the most pervasive ways of coercing or coaxing us to toe the mark is the media's use of the phrase "studies have shown." "They," who are presumably experts, say it, so it must be so. Even when we can see that a study used poor methodology, or it is being reported out of context, or it flatly contradicts other similar studies, they *have our attention*. We don't need to think because others are doing it for us, calculating what we need and want and putting it right out there in front of us (at a price, of course). There is enough variety to make us feel like unique individuals, all within the range of what is decreed and held as generally acceptable. Heidegger said of this trend almost fifty years ago that "calculative thinking may someday come to be accepted and perceived as the *only* way of thinking," which pushes the danger of falling into standing reserve to an even sharper extreme, because we would then seem to have no way out (DT 56). If calculative thinking were indeed to assume total dominance, we could be so narrowed, so reduced, that we would not just be stock but mindless stock, incapable of questioning, of breaking out of the comforting rut of uniformity and one-track thinking. Heidegger's use of the phrase "one track" makes one think of train tracks, which is no accident, because the image also points to the dominance of enframing—the way that techno-calculative thinking holds sway—as the understanding of being that shapes how beings are conceived and appear to us now. This is not the complaint of a curmudgeon or just a critique of culture. Heidegger is, in fact, careful to point out that our being in this situation is not due to laziness or some other personal failing on our part. It is the culmination of tendencies shaped over two millennia of Western thinking, and what Heidegger says about it emerges from the thinking of the first and other beginning. That is what Heidegger means, for example, when he says things like this about calculative, one-track thinking: "It reduces everything to a univocity of concepts and specifications the precision of which not only corresponds to, but has the *same essential origin* as, the precision of technological process" (WHD 56–58/WCT 32–34, emphasis mine).

So what do we do with this? I hope it is clear that Heidegger is not saying any of this to demonize technology or science. He even says that calculative thinking is at times necessary. It bears remembering, again and again, the way that all of this discussion takes place as thinking in the crossing of the first and other beginning. All of what is discussed here in "Memorial Address" and *What Is Called Thinking?* shows that the culmination of the history of metaphysics in enframing (and the possibility of opening another way of thinking at this extremity of thinking and human action) is no mere philosophical abstraction. It also shows that breaking through to a different way of thinking is going to take more than an act of will or a change of attitude. Neither is it saying or implying that the possibility of transformative thinking, preparing to leap into the thinking of be-ing, will "solve all these problems." This is more than a set of problems, and, in any case, problem solving is what calculative thinking does best (or at least does endlessly).

The leap to an other, noncalculative way of thinking can and does have practical import (a subject for chapters 4 and 6 of this book), but even before we begin to explore how it can be done, Heidegger warns us against either overestimating or underestimating thinking. He says it will not give us scientific knowledge or "usable practical wisdom." It will not solve cosmic riddles or give us immediate answers to our questions or problems. On the other hand, if it takes place that thinking is transformed to the thinking of be-ing, a radically transformed way of thinking, we can hardly underestimate "thinking's power for grounding . . . time-space" (WHD 161/WCT 169; GA 65: 60/CP 42). But that leaps so far ahead as to be no more than a barely intelligible hint of what is to come. For now, note well that the calculative thinking, one-track thinking, and uniform views described above are all enacted in language and dramatically reveal language's power to shape thinking and action. That we are beginning to see this and to think it as concretely manifesting abandonment of being places us already under way within the thinking of the first and other beginning and on the way to a transformative experience with language. However, it becomes necessary now to seek a bit more guidance from Heidegger on just how this thinking may be carried forward. We need more clarity about just what it is we are attempting here. Gaining that clarity in terms of understanding how thinking is a transformative experience with language through engaging language's way-making is the subject of the rest of this chapter. The simply stated but radically profound insight at the end of "Memorial Address" gives this exploration some structure. There Heidegger suggests that this noncalculative thinking can be characterized in two ways or by two bearings that he calls "releasement toward things" and "openness to mystery." These show up in many ways throughout the rest of this book, in addition to moving us forward into this exploration of language and thinking.

RELEASEMENT TOWARD THINGS, ON THE WAY

What Heidegger actually says in "Memorial Address" is that releasement toward things is meant as letting go of our attachment to the wonders of technology; such letting go, or being able to say "both yes and no" toward such things, is a necessary step in disentangling us from the pervasive trap of one-track thinking. This letting go in connection with concrete things (and not just technological things) is taken up as a major topic of chapter 4. Here it seems appropriate to note that there are also some philosophical "things" that we need to release if we are to learn the kind of thinking suitable for attempting to think nonreifiable be-ing (and the whole domain that opens up with it). I already alluded to this in a preliminary way by pointing out what Heidegger says about the necessity of unlearning what thinking already is if we are to learn what thinking might become. So far I have given some indication that the predominantly calculative mode of thinking shapes us so decisively that we can barely imagine how we might let go of it. So now it is necessary to take a closer look at some of the key features of our usual manner of thinking, all of which have led toward and move within enframing.

First, consider the notion of method. The use of the phenomenological method in the transitional thinking of *Being and Time* was a necessary aid to prepare for engaging more directly with transformative thinking, for what Heidegger calls a *leap* into an other beginning. Again and again, though, we need to remind ourselves that this thinking of the first and other beginning is not a linear, cause-and-effect, preparation-and-outcome kind of process (GA 65: 447/CP 314). We have seen already that *Being and Time,* though preparatory, is more than just that. To carefully accompany Heidegger through the thinking that unfolds there is already transformative. To even begin to suspect that "being" is a linguistic phenomenon rather than an actually existing item in an ontology is certainly a huge change in our usual way of thinking. And so it is with all of Heidegger's works. They are all more or less preparatory, with some also having more of the character of a leap into a radically different way of thinking in their attempt to bring "be-ing itself" to language. Considering the way thinking is *attuned* is helpful here. In chapter 1 we saw that the thinking of the first beginning was attuned by wonder at the realization that those thinkers did not know how to conceive or say the being of beings. There they are, beings, showing themselves to us, but what is this, really, this beingness of beings? Amazing, wonderful, mysterious, and yet there they are, right there in front of us! In retrospect, the word "wonder" is indeed the apt name for this, the attuning of the first beginning of Western thinking. What about the attuning of the prospective other beginning, which already begins to emerge in the thinking of the first beginning as such, in its decisive beginning character?

> *The grounding-attunement of thinking in the other beginning* resonates in the attunings that can only be named in a distant way, as
>
> $\left.\begin{array}{l} \textit{startled dismay} \\ \textit{reservedness} \ldots \end{array}\right\}$ intimating
>
> *deep awe . . .*
>
> The inner relation among these will be experienced only by thinking through the individual joinings. . . . There is no word for the onefold of these attunements. (GA 65: 14/CP 11; see also GA 65: 22, 395–96/CP 16, 279)

We have already encountered the first of these, startled dismay, which arises from the first genuine awareness of abandonment of being. The realization that "being" may not at all be what we have for so long thought it was or, in fact, "be" at all is startling and even shocking. The German word Heidegger uses here, *Erschrecken,* intimates something of this even to native English speakers. This shock jolts thinking into engagement with something that can barely even be thought or brought to language, with the be-ing of beings, which instead of offering to serve as ground manifests in and as ab-ground, the staying away of ground. The "deep awe" mentioned by Heidegger here arises more so the further along we go in this attempt to think be-ing. It will come into play in various ways later on. Here, however, we need to look closely at "reservedness," which plays out in and marks every turn of this thinking.

Already, when Heidegger tells us that there is no one word for the grounding attuning of the other beginning, reservedness is at work. How could there be one name for something that we are in the midst of, especially when it is only just beginning to emerge? We can name the attuning of the first beginning as wonder, with over two millennia of its playing forth behind (and in and in front of) us. How the attuning of the transformation of an other beginning plays out is only now starting to unfold. So the reservedness that is one mark of the attuning of an other beginning makes us cautious about being too quick to attempt to characterize the nature of this other beginning. It also reinforces what I said about the need to unlearn what thinking has traditionally been, that is, to let go of the compulsion to force this thinking into the patterns enforced by the methods and standards inherited from Western philosophy. There is much that could be said about this, some of which will only make sense later, after the discussion of time-space in chapter 3. (See, e.g., GA 65: 21–22/CP 16, where Heidegger says that "the manifold names do not deny the onefoldness of this grounding-attunement; they only point to the ungraspable of all that is simple in the onefold. . . . It traverses and thoroughly takes stock of the whole of temporality: the free-play of the time-space of the t/here.") Here, though, we need to be clear about at least this much: (1) instead of method, we are getting under way and staying *on the way* toward and within a transformative experience with language, and (2) instead of system and theory and their building blocks—concepts and representations—we have joinings and saying, which themselves remain open and dynamic, keeping thinking on the way rather than stopping, satisfied with some definitive outcome. That calls for more explanation.

"This preparation does not consist in acquiring preliminary knowledge as the basis for later disclosure of actual knowledge. Rather, *preparation* is here: opening the way, yielding to the way—essentially, attuning. . . . But the pathway of this enthinking of be-ing does not yet have the firm line on the map. The territory first comes to be *through the pathway* and is unknown and unreckonable at every stage of the way" (GA 65: 86/CP 60).

Whenever Heidegger discusses thinking, the notion of *way* comes into play, whether explicitly or implicitly. *Being and Time* itself was an effort to set out on the way toward preparing to ask the question of the meaning of being without knowing in advance whether the way was the only or even the right one (GA 2: 576/BT 437). *Contributions* reemphasizes the necessity of thinking's getting under way into the attempt to think be-ing. "From where else does thinking get its bearings, if not from out of the truth of be-ing? Hence be-ing can no longer be thought in terms of beings, but must be more deeply thought from out of be-ing" (GA 65: 7 my translation). But how can thinking get its bearings from something so strange? Asked another way, How can thinking move through our language of being(s) in such a way that it moves with and to be-ing? How can thinking go its way such that the way *is* the going?

Here the way of thinking is itself held in question while thinking moves along . . . on the way. "Way" carries many meanings. A way can be a road or path, a course, a direction, a manner, a method or means. What sort of way is it that thinking follows and on which it remains? Heidegger is very clear that by "way" he does not mean "method," nor is reflection on the nature of this way even a methodological consideration (GA 12: 168/WL 75). From Descartes forward (as outlined with particular clarity in part 2 of his *Discourse on Method*), method has been bound up with the quest for rationally determined certainty, enacted between subject (mind) and object, (my body and bodies in general). Epistemology—setting standards for method—follows metaphysical assumptions. In turn, method determines the manner in which something will be studied or investigated. This then also, in modern science and even more rigidly in our contemporary scientific-technical age, predetermines what counts as a result of thinking or investigation.[2] On-the-way thinking, on the other hand, is attuned by and to the matter for thinking, that which calls forth the questions that set thinking on its way. The region through which such thinking moves only becomes such in going along the way. It is not already there to be mapped out ahead of time or even to be discovered and then mapped out (be-ing as ab-ground is not *there* in any sense in which we usually understand that word). But neither is this just a matter of free, arbitrary invention. The paths of thinking open *and* follow the terrain. The terrain, however, is not only unfamiliar but also incalculable, moving in(to) ab-ground. The matter for thought is most decidedly not a problem to be solved. There is nothing fixed or provable in the outcome, much less anything certain. It is not even particularly appropriate to speak of there being an outcome, as if thinking were a means of going from one theoretical or practical point to another. The very notions of "outcome" or "conclusion" are not in the least helpful. Nevertheless, each step of the way is careful and precise (which is not to say, either, that it could not go in other directions, other ways) (GA 12: 167–68/WL 74; GA 65: 86/CP 60).

This emphasis on movement is no more intended to encourage the notion of movement for the sake of movement than it is an impulse toward certainty. The matter of thinking, for which one word is be-ing, reveals and hides itself in vibrating tensions that hold open possibilities: closure and dis-closure, sameness and difference (gathering and dif-fering), and showing and not-showing. Resolution, coming to rest on one side or the other, even temporarily, would only serve to close off possibility. The only "lasting element in thinking is the way" (GA 12: 94/WL 12). And it is not a linear progression, moving "from A to B." What then?

Heidegger names the movement of thinking in a way that at first may seem strange: *Erfahrung*, experience. Thinking undergoes, says Heidegger, a "transformative experience with language" (GA 12: 185–87, 239/WL 192, 119). How does one "experience" the movement of thinking? In German there are two words that both come into English as "experience," *Erlebnis* and *Erfahrung* (nouns, with the verb forms being *erleben* and *erfahren*). These both figure in Heidegger, but he uses only *Erfahrung* to name the kind of experience that is on the way in transformative thinking. Sorting out the distinction Heidegger establishes and is careful to maintain between these two words helps us understand what is meant by thinking as a trans-formative experience with language.

Erlebnis and *erleben* are both linked quite closely with the German word for life, *Leben*. Hence, the words carry the connotation of lived personal experience, or living through some occurrence or event. *Erlebnis* can even mean an adventure. Heidegger quite often uses this word in a context where he is talking about thinking and action, or life at the extremity of the history of metaphysics, in (unrecognized) abandonment by being, under the rule of enframing. "*Lived experience* corresponds to *machination*," he says, rather bluntly. He does not mean that lived experience is identical with machination but that the two are intimately intertwined. They enact our received obsessive drive toward the deceptive comfort of certainty, of cor-rectness and calculable explainability. Our everyday lives revolve around this compulsion, although it is in the deep background. The experiences we have foster our sense of our own existence as the subjects of experience in a world divided into such subjects and the objects of their experience.

"Lived experiences" take place in a dualistically framed world (GA 65: 132/ CP 92; GA 12: 122/WL 35). This understanding of lived experiences helps us understand even more clearly the ways in which this compulsion to cling to the results of calculative thinking affect us. Grasping for the comfort of stable correctness, with a lurking unease or unsatisfactoriness, a sense that we never quite have what we want, leads us to avoid some things and grasp at others. We grasp at group identity (being often deeply suspicious not only of uniqueness but even of those who prefer or claim for themselves a modicum of solitude). We grasp at shared ideas, at thoughts and values and value-judgments held in common. We foster the rule of "the experts" to tell us what we should think and believe and accept and do. Style and fashion give us a sense of change, of "something new," while at the same time keeping us in step with everyone else. Awe, reticence, solitude, stillness, and waiting seem decidedly out of place in a world characterized by problem solving, taking action (we simply must *do* something both as individuals and even more often in organized groups). Excitement lets us know that we are alive. We see the effects of what Heidegger calls "acceleration" in the sense that everything seems speeded up. This is not only in the obvious realm of technology (e.g., ever-increasing computer processor speeds) but also in the ways we spend our days. Not only do we have to be doing something at all times (even relaxation and meditation are now advocated as a way to increase efficiency and productivity or, at the very least, to acquire "peak experiences"), but we are bombarded by the constant flow of bits of information that I described earlier. This speeded-up character of our lives is, in fact, a significant hindrance to thinking (in the sense in which we are working toward it) in multiple ways. *Thinking* is not the acquisition of information or of lived experiences.

On the most down-to-earth practical level, just being able to *take the time* to slow down and think is incredibly difficult. When my students read "Memorial Address" for the first time this part invariably strikes home with most of them, even the ones who are not the least interested in philosophy: "Thoughtlessness is an uncanny visitor who comes and goes everywhere in today's world. For nowadays we take in everything in the quickest and cheapest way, only to forget it just as quickly, instantly" (DT 45). I ask them where they see this in their own lives, and I quickly receive plenty of examples, from toddlers' cartoon shows (which accustom the young to constant excitement and foster a very short attention span), to cramming for exams, television commercials with their brash and jumpy quickness, news shows with their thirty-second segments, and task and assignment

deadlines that hardly ever seem to allow us enough time to actually ponder something but rather force us to undertake everything as a problem
to be solved so we can move quickly on to the next task. Information is a
commodity that, as they say, goes in one ear and out the other. Information is at the same time something *to be experienced*. In spite of all this
quickness and busy-ness it is remarkable how passive the entire process is
and how little initiative it actually calls for. We are so accustomed to this
barrage of information, with its conflation of learning and thinking with
information acquisition and of all that with entertainment, that we hardly
notice it. In the conflict with Iraq in 2003 media reporters went in with
the invading ground troops to give added excitement to the actually quite
incoherent reportage of what was happening (much less why it was happening). They certainly did not show any of the dead Iraqi children whom
they surely saw. We are so continuously bombarded by "news" and by all
sorts of information that it becomes very nearly meaningless. This seeming openness, ironically, can function as a kind of censorship—a censorship not of information but of *thinking*. The following comment needs to
be heard:

> All of the details are utterly public. . . . The genius of the new censorship is
> that it works through the obscenity of absolute openness. . . . The betrayal
> of public trust is a daily story manipulated by the media within the narrow
> confines of "scandal," when in fact it's all part of the daily routine and every-
> one knows it. The media makes pornography of the collective guilt of our
> politicians and business leaders. . . . We then consume it, mostly passively,
> because it is indistinguishable from our "entertainment." What genius to have
> a system that allows you to behave badly, be exposed for it, and then have
> the sin recouped by the system as a sellable commodity![3]

As Heidegger put it quite a while ago, the refusal to accept any limits or
to feel any shame or embarrassment and certainly no awe at life or death
characterizes both machination and experience lived on its terms, with the
result that *thinking* is supplanted by "exaggeration and uproar and blind
and empty yelling" (GA 65: 131/CP 91). Again, this is not a "critique of culture" in the manner of one of the popular pundits. Instead, it tells us in
yet another way that the call of thinking in the crossing of the first and
other beginning is no mere philosophical abstraction. We are living in the
midst of enframing and of the machination that both flags and hides
abandonment of being. The upshot is a disempowering of language, an

impoverishment of thinking and also of our own range of possibilities, unless we can break through to another way of experiencing what comes to us by way of language. And this is precisely where we encounter the thought of a "transformative experience [*Erfahrung*] with language" in Heidegger.

Heidegger often hyphenates *er-fahren* and *Er-fahrung*, making more obvious the sense of travel or journeying that the word carries (*fahren* means "to travel"). To experience in this sense is to be under way, in motion, on a journey into the heart of the matter. Toward what? Toward the *Er-fahrung* itself, which is inseparable from the matter. This sounds circular, and on the surface it is, but not in the sense of going round and round, always returning to an identical place. The movement of the thinking experience is neither linear nor circular. It is not linear in that it is not a progressive sequence of statements leading to some conclusion or conclusions. It is not circular in that it does not simply retrace the same path. Heidegger says that the way of thinking is *transformative* (and thus is not circular) and yet "leads us only to where we already are" (and thus is not linear). In contrast to lived experience, which is engaged in grasping for certainty and similarity, in having and doing and being, transformative experience is open, ungraspable (if anything, in terms of the experience of attuning, it has us), and not characterizable as either active or passive.

How can a thinking experience lead us to where we already are and yet in that very process bring a decisive change? "Where we already are, we are in such a way that at the same time we are not there, because we ourselves have not yet properly reached what concerns our being" (GA 12: 188/ WL 93). In the first place, we are, we cannot help but be, speakers of language. That is where we already are. Then how can we not be there? Recall the beginning of *Being and Time:* we all know what "is" means, using some form of the word more or less correctly nearly every time we open our mouths. Yet we are so far from wonder at the presencing of things that we require a tremendous effort of thinking just to prepare to ask the question of the meaning of being, much less begin to think be-ing's holding-sway. Similarly (and not accidentally so), we talk and talk, but we do not wonder at language itself. A deep question concerning language, like the question of the meaning of being, is not something that can be formulated in one interrogative sentence. It gradually emerges and is experienced as questionable only "in the light of what happens . . . on the way" (GA 12: 229/WL 111). A transformative experience with language shifts us back to where we already are (as speakers of the "is") but in a way that opens up the possibility of wondering and asking about what is ownmost to language itself and to its creative energy. It opens the possibility of being attuned to (and by) an opening for the thinking of be-ing. But *how* is one to think, how is one to even *begin* to do this? We already know that Heidegger is not going to give us some method or a map that lays out the way. Talking about thinking is no substitute for thinking; the way must be *traveled* rather than explained. It must be experienced, and that is the task of much of this book, to travel with Heidegger along some interrelated paths of thinking and to follow through by exploring the transformative possibilities that may open up along the way. But that doesn't sidestep the need to have some indication of how to proceed. We know that letting go of the rule of method and of the idea that language is some kind of commodity for acquiring information are important steps at the beginning. These are some aspects of the releasement toward things mentioned in "Memorial Address" as a way to begin thinking. Now it is time to turn to the other clue he gave there: openness to mystery. It serves as a guideword (*Leitwort*) to other keys to getting under way on the path of transformative thinking. It will also help us see more clearly the nature of some of the other philosophical "things" that we need to release. So next we will engage in some reflection on openness to mystery in relation to (1) withdrawing and questioning, (2) guidewords, and (3) the nature or ownmost dynamic of language.

Openness to Mystery: Withdrawing and Questioning

In shifting from the thought of being as ground to attempting to think be-ing as ab-ground, we certainly encounter something that seems mysterious. Reservedness about saying more comes rather easily at this point! But the reservedness that is one facet of the attuning of the other beginning has more depth and energy and power than that would indicate. It reflects to us something of the dynamic of be-ing and ab-ground. *Being withdraws.* It withdraws from any direct perception or conceivability. But, as Heidegger puts it, this refusal to be grasped by the old ways and means "is the foremost and utmost gifting of be-ing" and its originary way of holding-sway (GA 65: 241/CP 170). The ab-ground-ing withdrawal of be-ing plays out decisively in the thinking of the first and other beginning, all the way back to the original impulse to think being, holding the being of beings in question. In the conceiving of being as grounding presence the dynamic of the withdrawing of be-ing is covertly at work, reflected in more than one facet of that thinking. The Greeks were drawn to wonder at beings as things *arising* (*physis*) and coming to presence to appear for our perceiving and conceiving. In conceiving their being as grounding presence the pres-enc*ing*, the aris*ing* continually withdraws from thinking; it remains un-thought and concealed within the metaphysical language that speaks of being and beings. Such language at one and the same time grants being and conceals presencing. This movement of revealing and concealing takes place within language. But *that* this is taking place in this way is also hid-den. The thought of being is taken as reflecting the way things really are, and so the role of language, *in such a framework*, must also be understood as expressing rather than creating being. In accord with this Aristotle understood language in general in this way: a being out there in the world is perceived, giving us the image and idea of that being, which is then expressed in language. So, early in the history of metaphysics, the dynamic of presencing, of be-ing, and of language's creative role in thinking them remained very deep in the background. Effectively, "it" (all of it) withdraws.

A word of caution is called for here. To describe the situation in this way could create the impression that if the Greeks had only been more care-ful, or a bit sharper, or less bound to what could be accomplished through *technē*, they might have been able to think and say be-ing itself. While we certainly cannot rule out that things could have been otherwise and that they might have persisted in attempting to think *physis*—arising itself—rather than the being of what had arisen, we cannot say that they could have grasped and conceptualized be-ing in a definitive or conclusive way. Why? Because the dynamic presenc-ing of be-ing withdraws, and this is not due to our—or the Greeks'—failure or inability to think or pay atten-tion. Instead, as Heidegger puts it, "that we are still not thinking stems from the fact that the thing itself that must be thought about turns away from [us], has turned away long ago . . . since the beginning. . . . It withdraws" (WHD 4–6/WCT 7–8). One might ask, as Heidegger does, "how can we have the least knowledge of something that withdraws from the beginning?" We can only because, as he said in *Contributions*, this withdrawing is a gifting. How so? This withdrawing is not a blank nothing, going into void nothingness. The more persistently we try to think this matter, the more elusive it becomes, while at the same time it becomes more and more clear that there is something happening, something at work. The capacity for being drawn and pulled along toward and into this matter is, asserts Hei-degger, what allows us to come into our own as humans. To let ourselves be reduced to mindless stock, only able to think calculatively, is to be less than that. Either way, we have language and the capacity to think. At stake is whether we are reduced to only a tiny fraction of our possibilities or we attempt to go farther and deeper and risk being decisively transformed as we do (WHD 5, 51/WCT 9, 17).

In being open to this possibility, in being open to its mysterious pull on us, we are heeding something in us, something that tells us (even if only by way of hinting and by eluding our intellectual grasp) that be-ing is not something separate from us. Somehow, it—*in* its very withdrawing—speaks to us at a very deep level. That is why Heidegger could say to his neighbors that, to get under way in thinking, it is enough to genuinely heed and ponder what is of concern to us, what is close to us. Whether this call to think the mystery of be-ing is provoked by pulling a perfect scallion from the soil, by being plunged into darkness when the power goes out, or by trying to understand how the history of Western philosophy shapes us, it draws us into (the) be-ing that is not just of "beings" in some vague and general sense but is unique in each instance and pertains to us and everything of concern to us. This dynamic, relational enowning and arising of everything is so near to us that it could be said—as Heidegger does say—of it that it is our very heart's core. But as close as this is it calls us to *think*, because the more we are touched and moved and drawn by it, the more elusively compelling it becomes (WHD 5, 51/WCT 9, 17; DT 47).

Being drawn on by what thus withdraws is what sets us on the way of thinking and inclines us to continue moving along the way. It compels and attunes our questioning. Over and over Heidegger says that what is worthy of questioning is what is worthy of thought, of attempting to bring to language "the inconspicuousness of a withholding, its riches" (GA 12: 227, marginal note). The questioning that is intrinsic to thinking that always remains on the way does not ask for answers. It places "questions that seek what no inventiveness can find" because the matter itself is inexhaustible and utterly resistant to calculative thinking. To persist in this thinking is not to persist in the attempt to find an answer but rather to *persist in questioning*, to remain in motion, in unresolved tension and openness to transformative possibility. The questioning that holds thinking on the way is in no way a matter of question and answer (distinct moments in a linear process). Rather, being drawn on by withdrawal into ab-ground keeps thinking ever more deeply, even compellingly, engaged with the matter. This unfolds as a different experience of questioning, jarring our presuppositions about what questions and answers are. "If an answer could be given it would consist in a transformation of thinking, not in a propositional statement about a matter at stake." And such a radical transformation of thinking also transforms us, whose very nature is in question and at stake, as this thinking proceeds (WHD 5, 81/WCT 8, 116; GA 65: 85/CP 59; TB 55).

Withdrawing and the questioning it evokes are in oscillating play, pulling thinking along, opening a way into the matter. Going the way of this interplay calls forth, is called forth by, and enacts an attuning. What does this say? In the first place, it indicates that neither thoughtful questioning nor attuning causes one another. Instead, there is a resonance between questioning and attuning whereby the attuning both compels and responds to the questioning and the ever-deepening questioning calls forth a response that in turn attunes further questioning. The questioning responds to the beckoning hints and traces and echoes of be-ing's withdrawing and in turn "challenges be-ing to thoroughly attune the questioning" in its very withdrawing (GA 65: 86/CP 60). So it is that thinking takes its bearings from be-ing while opening the region of the thinking of be-ing and getting under way within it.

Earlier, I briefly mentioned some of the words Heidegger uses to name the attuning that is in play within this thinking of the first and other beginning: reservedness, dismayed shock, deep respect or awe, and intimating. To what has been said earlier of startled dismay and reservedness I can now add a better understanding of the way that intimation works here. Withdrawal of be-ing, evident in a growing sense of the shakiness and uncertainty of grounding (abandonment of being), is not simply obvious but yields intimations that evoke dismayed awe, deepening into an encounter with be-ing's own reservedness (withdrawal, refusal of grounding, denial of any answers, etc.). This attunes a questioning that is carefully reticent, letting what is reserved be held back, yet being persistently attentive to the hints and intimations that come forth in the course of the thinking. Reservedness pervades all facets of this attuning. Perhaps that is why, of all the names for the attuning of the other beginning, reservedness is consistently called *grounding* attuning: it pervades thinking as both the withdrawal of be-ing and the manner in which transformative thinking unfolds (GA 65: 22–23, 33–36, 80/CP 16–17, 24–26, 55–56).

It unfolds, of course, in language, emerging from what has already been spoken (thought) and thus has been granted to us (GA 12: 164–65/WL 71). We are on the way within language. We ask questions of language concerning what language reveals (that which has been spoken and which is being spoken) and what remains concealed (and unthought) in what is said. But what, precisely, has been spoken and thus granted to us? In section 44 of *Being and Time*, in a discussion of the hidden meaning in the words *logos* and *aletheia*, Heidegger says that "the ultimate business of philosophy is to preserve the force of the most elemental words in which Dasein experiences itself" (GA 2: 291/BT 262). In the history of Western philosophy many of these elemental words are "words of being," naming the ways in which beings have been determined in their ways of presencing, thereby also determining human experience of beings and self. Others emerge first, or with altered meanings, in the thinking of the first and other beginning, in its attempt to say the arising insights that begin to emerge. To heed them in our questioning is to seek hints of how language arises and holds sway, holding us together with beings in ways that shape our own arising and enduring. To begin to understand this also helps immensely in understanding Heidegger and in actually *thinking* after Heidegger.

Guidewords: Releasement Toward Things, in Openness to Mystery

257. Be-ing.
Here lie the boulders of a quarry, in which primal rock is broken:
Thinking. . . .
Being and the difference to a being.
Projecting be-ing open.
En-thinking of be-ing.
Essential swaying [*Wesung*] of be-ing.
History.
Da-sein.
Language and saying.
"A being."
The question of crossing (Why are there beings at all and not rather
    nothing?) . . .
The incalculable.

(GA 65: 421/CP 297)

Ur-rocks from the same quarry. This is another way of saying: joinings from within the thinking of be-ing. Whether these Ur-rocks are a word (being, be-ing, Da-sein), a phrase, or a full sentence (as the "Why . . ." near the end of the list), they are all guidewords for thinking. In chapter 1 I suggested that it is helpful to see the words and works of Heidegger as joinings, working together in the same way as the joinings within *Contributions*. They all speak from and into various facets of the thinking of the first and other beginning, echoing and resonating with each other in a way that continually deepens and enriches and moves thoughtful questioning forward. These guidewords, coming from the "same quarry" and providing much of the dynamic energy of Heidegger's works, function in much the same way.

These guidewords emerge in the play of the historical thinking of being, resonating as words that in one way or another echo the holding-sway of be-ing. This is not and never could be a complete list, any more than *Contributions* is the "last word" for thinking. But neither is it likely that any of these would be exhausted and somehow "lose their place," be ground to powder in the rock-crusher and blow away as dust on the wind. As thinking continues, other words in other languages will no doubt serve as guidewords for this thinking. Heidegger himself had occasion to consider at least two of these, *koto* (Japanese) and Tao (Chinese) (GA 12: 135–36, 187/WL 45–47, 92). This is not to say that these words—and others not considered by Heidegger such as *sunyata* (Sanskrit) and *rigpa* (Tibetan)—say "be-ing" in precisely the same way as do the Greek or German or English words. Neither is it to say they do not. What I am suggesting is that there are radically different linguistic and philosophical traditions that "say the same" in the way that Heidegger uses that word: belonging together in their difference, they somehow echo be-ing's play of revealing and concealing, arising and withdrawing, gathering and dif-fering.

"Being," I have said, manifests in and only in language. To that, Heidegger adds that "every saying of be-ing holds itself in words and naming" (GA 65: 83/CP 58). But be-ing is not—and cannot be—said directly in a name or a bushel full of names. Words of being—not only philosophical terms but also the everyday words and meanings by which being is understood and beings interacted with—both say (show) *and* obscure, they both reveal *and* conceal. And I say that about being and beings—never mind be-ing! Thinking must move with this contending interplay of revealing and concealing. Within that interplay other tensions resonate productively. One is between the static character of the language of being (especially the philosophical concepts and their offspring) and the ever-moving sway of be-ing. Another is between the sense of ground and rest involved in the language of being and ab-ground's unfathomable regioning of the dynamic holding-sway of be-ing. If attended to carefully, these tensions do not block thinking but, on the contrary, help move it forward. Heidegger suggests that thinking must accommodate the common meaning and go a certain stretch of the way with it, persisting in attuned thinking through being open, heedful, questioning, reserved, and reticent and letting go of the expectation of a decisive answer (GA 65, 83–85/CP 58–59).

In carefully thinking through and becoming more aware of what is said (and hidden or even closed off) in the common and philosophical meanings of key words and ideas, their rigidity is loosened. They are opened up as thinking itself stays open to what is and can be newly said (fresh meaning in "old words," new ways of thinking and interpreting what is said). In carrying out the thinking that lets elemental guidewords open the way, both of the directives from "Memorial Address"—releasement toward things and openness to mystery—are crucial. In the remainder of this chapter *releasement* involves letting go of the notion of concepts as we usually understand them. Openness to mystery comes into play as we see some ways to work with the hinting, intimating dynamic of language and then also as we try to understand what is ownmost to language, that it has this power to reveal, to conceal, and to transform us as we think.

Guidewords are not concepts. Two questions arise. Why not? And how are they to be thought at all, if not in this way?

To understand Heidegger's response to these questions we need to have clearly in mind what we mean by "concept." Strictly speaking, a concept is a word that, in set or permanent form, captures a general or universal representation of some class of entities. Concept and representation are thus closely intertwined notions. "Representation" at root means "substitution." One thing stands for another, presenting a reproduction of its likeness. "Concept" implies fixity, permanence, universality. "Concept" comes from the Latin *capere*, "to capture." The German word for concept, *Begriff*, comes from *begreifen*, "to grasp." To conceptualize is to attempt to capture something in a word, to grasp and hold it available in constant presence. "Representation," re-present-ation, involves substitution and implies a hierarchical dichotomy between the representation and what the representation is of. The implication is that there is something really present to be re-presented.

How can I say that guidewords are not to be thought as concepts or representations? Are not *all* words in some sense concepts that represent something else? These questions speak from within our traditional view of language, going back at least to Aristotle's *On Interpretation*, where an operation of substitution or representation is shown to be fundamental to the speaking and writing of language. Written words stand for spoken words, which stand for thoughts, which stand for things. Words are thus held to be concepts that grasp, fix, and hold these substitutions (representations) in place for our common comprehension, making communication possible. For the most part this remains an accurate description of how a great deal of language function in both everyday conversation and thinking.

What needs to be held in question here is the role of concepts within the kind of thinking opened up by Heidegger. Thinking that remains *on the way* moves within an ongoing play of revealing and concealing, carried on in various levels of the language. Concepts limit and fix the meaning of words, holding it in place for our consideration. Through this holding concepts participate in revealing and disclosure. But concepts also necessarily serve to mark out a certain closure. In presenting a definitive meaning they tend to halt further questioning. But if one *persists* in questioning and thinking, this very limitation can also give hints of an overflow of meaning that is not disclosed and that withdraws from superficial thinking. Concepts in and of themselves are thus not inimical to thinking. "Even where thinking is in a certain sense concept-less . . . the metaphysical manner of forming ideas is in a certain respect unavoidable" (GA 12: 110/WL 25). It is not that we need a change in linguistic expression, that we need words that somehow "are not concepts." Rather, what is called for is a change in how we hear and think these words. The key here is to keep in mind that if we are to remain open to the possibility of a transformative experience with language, the tension between the disclosing and the closing functions of concepts must remain unresolved. This requires wariness of several pitfalls, any of which could derail thinking from its way, shunting it into the dead end of rigid closure or an utterly open (but impossible) expanse of pure disclosure. To conceptualize is to grasp. The question under consideration concerns *how* we are to grasp something and yet remain on the way in thinking. We need to grasp things lightly in a way that avoids doing violence to what is being thought.

This potential violence or violation is due to the way that concept formation is embedded in traditional metaphysics and epistemology. "Modern philosophy experiences beings as objects. It is through and for perception that the object comes to be a 'standing against.' As Leibniz clearly saw, *percipere* is like an appetite which seeks out the particular being and attacks it, in order to grasp it and wholly subsume it under a concept, relating this being's presence back to the *percipere* (*repraesentare*). *Representatio*, representation, is defined as the perceptive self-representation (to the self as ego) of what appears" (GA 7: 240/EGT 82). If we see ourselves as subjects and things as objects, then the relationship between the two is, as the word "object" (*Gegenstand*) implies, one of standing over against. This sets the stage for the aggressive grasping of conceptualizing. The concept is established as representing (imitating, standing for, substituting for) the object that has appeared to the subject's perception. If this representation is seen as the rigid determination of some reality ("wholly subsuming" the appearing thing under the concept), it does violence to the "object" as the thing it is, which is always more than can be grasped in this way. The overflow, the undisclosed meaningfulness that provokes wonder, questioning, and thus *thinking*, is effaced. Therefore, if we take concepts as occurring within and further cementing rigid categories such as subject and object, substance and attribute, or reality and unreality, we will not take note of the intimation of what withdraws from thinking, and our thinking will have lost its way, ceasing to be on the way.

But there seems to be a dilemma here. We do not want to be limited to the traditional categories and ways of thinking, but surely there is *some* kind of "grasping" involved in *any* thinking. Yes, there is. Heidegger himself said that even in this concept-less thinking we cannot completely avoid making use of grasping and conceiving. In the first place (as I already discussed), one of the major matters to be thought (and this includes some significant guidewords) is the complex of words that have traditionally named being. Heidegger says of those that we need to understand what was (and is) grasped under those names, going along a stretch of the way of thinking *with* them, because only then will the overflow and the unthought begin to emerge for further thinking (GA 12: 109–10/WL 24–25; GA 65: 83–84/CP 58).

I mentioned earlier that one challenge to staying on the way is that our very grammar arranges what we say in line with those metaphysical categories. Therefore, thinking tends to move toward closure almost as soon as it begins. The particular value of what Heidegger calls guidewords is that they are words that carry such an unease, such an evocative, beckoning hint of what draws thinking along in its withdrawal, that to persist in thinking them tends more to opening than to closing or stopping. The longer we work with them, the less we find ourselves thinking in terms of subject, object, substance, or reality. To notice that is to notice that we are already undergoing a transformative experience with language.

To return more directly to the thinking of the first and other beginning, the twofold (of being and beings) *itself* is not at all an object in itself but appears to thought only as it has been represented, in the sway of usage, in the ways in which the "words of being" unfold in the history of language and thinking, in how they hold sway over the years. They reveal beings in certain ways while always concealing their own way of creatively holding-sway in so doing. This calls on us not to pay heed to such words as fixed concepts representing some object (being, beings, being *of* beings) but rather to attend to them *as used*, that is, in the ongoing movement of language. It calls on us to experience the sway of usage. Traces of the non-conceptual overflow appear as a faint unease in usage or as strange or per-plexing changes in usage that might be encountered in reflecting on the past sway of usage (etymology).

"The mere identifying of old and often obsolete meanings of terms, the snatching up of these meanings with the aim of using them in some new way, leads to nothing if not to arbitrariness. What counts, rather, is for us, in reliance on the early meaning of a word and its changes, to catch sight of the realm pertaining to the matter in question[,] . . . [the] realm . . . in which the matter named through the word moves" (GA 7: 42/QT 159). The key—yet again—is being on the way and undergoing an experience with language. Heidegger's ongoing engagement with etymology is not for the purpose of trying to change our current usage or for determining whether we understand a word "correctly." Changes that have taken place in the meaning and usage of elemental words reflect changes in the way language and things are experienced. To thoughtfully recall former meanings is to open up the possibility of experiencing language and things differently, not precisely as those of former times did but, rather, in a way that is altered by an openness to long-hidden possibilities. This reflection on changes in meaning and on the relationships between changing words may help us to glimpse a few hints of the matter in question and its movement in lan-guage. So reflection on etymology is one way in which thinking, engaged with guidewords, is able to remain in motion, to persist in questioning, to be drawn on by what withdraws.

This brings me back again to the question that led to that brief discus-sion: isn't there (whether we call these words concepts or not) some kind of grasping taking place? Yes, of course there is. It is not as if thinking were just creation ex nihilo or, at the other extreme, wandering endlessly in some vague fog, apprehending nothing at all. Heidegger addresses this issue, making a very fine distinction within two senses of the notion of concept and the grasping it involves. A crucial passage in *Contributions* helps us carefully think this through.

> What is grasped here—and what is always and only to be grasped—is be-ing in the joining of these jointures. The masterful knowing of this think-ing can never be grasped in a proposition. But what is to be known can just as little be entrusted to an indefinite and flickering representation.

Concept [*Begriff*] is here originarily the "in-grasping" [*Inbegriff*], and this is first and always related to the accompanying co-grasping of the turning in enowning. . . .

In-grasping here is never a comprehensive grasping in the sense of a species-oriented inclusiveness but rather the knowing awareness that comes out of in-abiding and brings the intimacy of the turning into the sheltering that lights up. (GA 65: 64–65/CP 45–46)

There are some things mentioned here (especially the "turning in enowning") that will be taken up and make more sense later, after the discussion of timing-spacing-thinging in chapter 3. For now, it is enough to say about that as-yet-unexplained matter: the turning in enowning is another way to say the dynamic of be-ing, saying something of how it brings all things into their own such that they are "sheltered" while brought to light. What, then, is the difference between conceptual grasping and in-grasping? In-grasping is attuned to and by *reservedness* and abides in awareness of ab-ground. This is not just in relation to be-ing and its dynamic (the turning in enowning) but also to beings. So Heidegger says that in-grasping is co-grasping of this dynamic, which always eludes a final or definitive grasp, holding itself in reserve, and in this elusive withdrawing it attunes us to a corresponding reservedness. In plain English, "be-ing itself" is always just out of thinking's reach, and yet, as we have already seen, it is not simply nothing. There "is" something happening, an energy that draws us, but anything we can say of it will only be partial and provisional, not comprehensive and definitive. We can in-grasp be-ing because "it" is not something separate from us. Here we must decisively let go of subject-object dualism. Be-ing is neither subject nor object; it is one word for the dynamic arising of everything and anything, including us (this is the main topic of chapter 3). So in "grasping" something that "rings true" or discloses a facet of this dynamic, we are not (only) grasping something "out there" (although it does pervade everything) but also "in here." At some point not only does the subject-object distinction fall to the side but so does the distinction between "inner" and "outer." In the language of *Being and Time*, which calls us Da-sein, we are the *Da*, the t/here for the disclosure of beings. T/here: not just here or there but *open* to the dynamic web of disclosive relation to beings. So in-grasping, unlike grasping, changes us, every time, unlike conceptualizing and theorizing, which leave us in charge at a carefully maintained objective distance (GA 65: 13–14/CP 10–11).

How can we foster such in-grasping? One way is by taking the Ur-rocks, the guidewords that arise on the path of thinking, as hints rather than as defining concepts. Hints suggest and provoke rather than declare or explain, beckoning thinking to move toward what they name without offering anything conclusive (concluding the movement of thinking). They guide thinking on its way, beckoning toward and evoking an experience of what they name. There is always an overflow over what is revealed, an overflow that withdraws from thinking's approach, leaving only an intimation of a path along which thinking could be drawn. Hints are akin to gestures. In a dialogue with a Japanese acquaintance Heidegger linked hinting, gesture, and bearing in a way that I find helpful. In Japanese Noh plays the stage is empty. The actors must convey a sense of the surroundings through subtle gestures such as a hand raised and held slightly above the eyes to indicate a mountainous landscape. Of course, this can become very conventionalized, but it need not be. The example draws from Heidegger the thought that such a hinting gesture is the "gathering of a bearing," and his friend adds, "Thus you call bearing or gesture: the gathering which originarily unites within itself what we bear to it and what it bears to us" (GA 12: 102/ WL 18–19). A hint, whether it manifests in language or in a physical gesture, intimates its meaning rather than spelling it out conclusively, and grasping the meaning depends on a *gathering* of what we bear to this encounter with the hint and what it bears to us. There is a relational context in play that, in its variability and dynamism, quite often disallows one conclusive meaning or interpretation. In English idiom occasionally someone will talk about a gesture being "pregnant with meaning" without, perhaps, realizing how apt this description is. "Gesture" and "gestate" are cognates, both going back to a root meaning of carrying (which, of course, also has a sense of "bearing"). Gestating and gesturing both carry more meaning than is immediately obvious. They show something, but much is hidden below the surface to emerge or be drawn out gradually. And so it is with the hinting nature of the guidewords for thinking.

One implication of this is that, since guidewords are provisional (and carry their meaning through the hinting gesture), they will always tend to be somewhat ambiguous or to have multiple meanings (WHD 68/WCT 71). They do not, in other words, narrow thinking down to one track but open a weave of several possible ways to go. This can be understood a bit better if we put it in play with something that at first sounds just the opposite: *thinking the same.*

In reading Heidegger one encounters many phrases that take the form of the following examples.

"Ereignis ereignet" [enowning enowns]
"die Sprache spricht" [the language speaks]
"die Stille stillt" [the still stills]
"die Gegend gegnet" [the region regions]
"die Nähe naht" [nearing nears]
"Be-wëgung be-wegt" [movement moves, way making makes ways]

These phrases almost invariably occur at crucial junctures in the text, elim-
inating the possibility that we could gloss over them as rhetorical embell-
ishments or as empty tautologies used simply for aesthetic effect. To think
along with Heidegger here requires insight into the way in which "thinking the same" moves into and with the matter for thinking.

Thinking takes us back to where we already are but in a way that we
have not been aware of before. It does not move in a linear sequence of
progressive steps but resonates dynamically in such a way that "forward"
and "back" are not particularly helpful descriptions. The steps of thinking,
whether they appear to be moving forward or back, are, says Heidegger,
coming together in a "gathering of the same" (GA 12: 197, my translation).
Two questions (or, perhaps, two ways of asking the same question) arise.
How is thinking that gathers itself on and to "the same" also transformative? How does such gathering of the same remain *on the way*?

When Heidegger speaks of this matter in his own language he always uses
the word *Selbe* (same), never *Gleiche* (equal or alike) or *Identische* (identical). Why? "The same is not the merely identical. In the merely identical,
difference disappears. In the same the difference appears, and appears all
the more pressingly" (ID 111/45). How are we to make sense of this? In everyday usage we tend to use "same" and "identical" as interchangeable syn-
onyms. But for Heidegger "same" means the *gathering* of what, although
differing, belongs *together*. This understanding of "same" is anything but
arbitrary. In its historical unfolding "same" does carry those senses of *gathering* and *together* that Heidegger stresses: *versammeln*, "to gather"; *zusam-
men*, "together." Both words trace back to the same root, which appears in
the German suffix *sam*. That was originally an independent word, a word
that is also the source of our English word "same."[4] The same here is what
gathers and holds different things into their belonging together. It does not
do so by way of some fixed center, whether as a unifying principle, necessary connection, mediating term, or even relation between two objects.
It is more like what we encounter over and over again on the way: an unresolved tension between differing and belonging together. The emphasis

is on the *belonging*, not on the "together" (which could just be the linking of two terms mediated by some fixed center) (ID 92–95/20–32). What is the same in this sense, that is, what is gathered into a belonging-together, can thus not be predetermined or conceptually grasped and fixed rigidly in place. The belonging *itself* withdraws from thinking, hinting at be-ing, at arising, at the dynamic relationality of enowning (timing-spacing-thinging, soon to be discussed in detail).

To think the same is to gather one's thinking to a place where what differs belongs together and belonging together does not obliterate difference. This avoids the tendency to think in terms of a difference and identity dichotomy, in which the movement between two differing thoughts is seen as oppositional and the temptation is to let that dichotomy resolve in favor of either identity or difference. Here, the multiple possibilities carried by the words are held open, preserving the power of elemental words, letting them serve as guidewords for thinking. Persistence in such thinking allows what is thought (held in question) to unfold its multiple possibilities from out of its own internal resonance, to be gathered in thought. There are at least two distinct ways that this thinking the same comes into play in Heidegger and in thinking after Heidegger. If we stay with the notion of the guidewords, this can be clarified without much difficulty. The guidewords are the same (gathering differences that belong together) both with(in) themselves and in play with each other (as multiple ways to gestate, gesture, and carry out the thinking of be-ing).

Guidewords join sameness and differing in themselves, one by one, in two ways. The most obvious is one that I have already mentioned: multiple meanings. The other is at work in the list of seeming tautologies that opened this segment of the discussion a few paragraphs ago. Most of those guidewords have yet to be discussed individually, but still, the point can be made. Enowning, stilling, nearing, way making, and so forth all *do* what they name. They cannot, therefore, be reduced to easily grasped nouns, beings. In each case this will also draw out more questions that move thinking deeper into the matter. How, for instance, does "the region" *region*? Where? Involving or including what? So thinking the same can open up the internal resonating of a guideword in its verbal dynamic, its way-making character, as well as its carrying multiple meanings.

Thinking the same also pertains to the belonging-together of all the guidewords that move the thinking of be-ing, the thinking of the first and other beginning. They are quite obviously different words, each with its own complex history and its own internal differing. Yet they *say the same* in saying some significant facet of the heart of the matter: be-ing. They gather the various aspects of the matter. They come together as joinings: echoing, resonating, reflecting, ringing and sounding and resounding in dynamic interplay with one another. If they are heard and thought in this way, transformation is inevitable.

Notice that *joinings* has come up several times in the discussion: joinings of the different sections of *Contributions* (from which I first drew the word), joinings of the works of Heidegger, and now joinings of the guidewords that move transformative thinking. In the East there is an image that can help us think how joinings function. It goes by the name of Indra's net. It is often used by Buddhists to help explain *pratitya samutpada* (interdependent arising). Indra's net calls on us to imagine an indeterminately large webwork of brilliant, highly polished, transparent jewels of many colors. Any facet in any jewel will reflect all its neighbors, and they in turn reflect all their neighbors, and on and on. If only we had eyes sharp enough, we could see that in reflecting its neighbors each jewel also reflects all of what its neighbors reflect, and on and on it goes, throughout the web. Thus each jewel reflects *all* the others while at the same time it is unique. Each jewel has its own color and shape and number of facets, and each facet of each jewel reflects all the others in a way that is uniquely its own, not identical with any other, since they all reflect from different places and different angles. The joinings we encounter in thinking are like this. They resonate and color and change each other while each says what is uniquely its to say. The joinings each say the thinking of being in a different way with a different bearing, gesturing-gestating a different "jewel" but resonating always to a conjoined questioning, saying the same, gathering what belongs together. Just so, says Heidegger of the guidewords in the thinking of be-ing, "they never name the essential sway of be-ing as properties but rather in each case the *whole* essential swaying of its essential sway" (GA 65: 486/ CP 342). Therefore, any one of the guidewords carries the others along with it, enabling us, if we can persist in thinking them, to move in many ways through the web of conjoined meanings.

So much has been said of guidewords' functioning transformatively that a recapitulation will probably be helpful. How do guidewords stand in distinction to concepts traditionally understood? Concepts grasp at being or at an idea of (a) being, thought within the framework of metaphysics, which means also (since the modern era) within the framework of subject-object dualism and (in our contemporary world) within techno-calculative thinking and its aim of univocal expression. All of this tends to close off rather than stay open to divergent possibilities. Concepts emerge from methodical thought and are assembled into theories and systems based on some foundation or ground. In conceptual thinking "same" means "identical." Heidegger refers to a definable concept as a "bucket with a sense content," in contrast to guidewords, which are like "wellsprings" that must be "dug up" and tended, awaiting what flows forth (WHD 89/WCT 130). Guidewords are not re-presentations (they are not aiming at being, at what is present) and do not construct objects from within subjectivity. Dualisms (being-nonbeing, subject-object) do not structure this thinking. Guidewords are not definable, and they have multiple meanings. Their ambiguity holds open the resonance of gathering and differing; the "same" here does not imply identity. Since they are not concepts they will not serve to construct theories. (How could they be *verified?*) "This thinking and the order it unfolds are outside the question of whether a system belongs to it or not," but this does not mean that it is arbitrary or chaotic (GA 65: 65/CP 45). Our thinking of these nonconceptual guidewords is attuned by the reservedness that accords with their moving within ab-ground (GA 65: 36/CP 26).

So it is that persistence in questioning, being drawn along by what withdraws, and attentiveness to the multifaceted guidance of elemental words all serve to keep thinking always on the way toward the possibility of a transformative thinking experience, a transformation that arises from within the interplay of the words themselves. Rather than being caught within closure, thinking thus remains within the movement of the unresolved tensions that hold possibility open: the tension of question and answering response, the tension of revealing and concealing (withdrawing), the tension between the surface meaning of a word and the possibilities that are carried along through an unease in the sway of usage and the tension of the self-gathering belonging-together of what differs.

To approach the matter for thinking in this way is to remain open to difference and possibility in the gathering of words, remaining open to what Heidegger calls "the strangeness which lies in the matter itself" (WHD 48/WCT 13). The strangeness in what is to be thought makes us un-easy, keeps us from coming to rest in a familiar place. As thinkers in the Western tradition our familiar places are rigid concepts, representations, and the larger structures built from them: definitions and theories. The thinking that remains always on the way will not come to rest in such familiar places but will move through them. Still not yet considered in depth, though it has been trying to rise to the surface, is this question: What is language, really,

that it can "do" all this? What is ownmost to language itself?[5] What is its
enabling dynamic, its power to both reveal and conceal, its capacity to both
open up and hide things of such depth and range?

Openness to Mystery: Saying and Way-making

Heidegger moves through one discussion of these questions by saying that
thinking (seriously and persistently questioning) what is ownmost to language (the way that language arises and endures *as* language) evokes an
experience of the language of the holding-sway of what brings things into
their own (be-ing as enowning). This is the way that arising itself arises,
emerges, discloses itself, and somehow *speaks to us* (GA 12: 166, 180–81, 189–
90/WL 72, 76, 94). How strange! What does this mean? Arising and holding-
sway itself *speaks?* And it is this *speaking* or disclosing that then emerges
in language? These questions cannot be sidestepped because here again we
find ourselves moving into the heart of the thinking of be-ing. Let's place
some guiding thoughts into joining and then try to think through what
it is that they open up.

> The emerging as such of language can be determined in no other way than
> through the naming of its origin [*Ursprung*]. One cannot, therefore, give out
> a definition of language's emerging as such and declare the question of its
> origin unanswerable. The question of origin includes the determining of the
> emerging as such of origin and of arising [*Entspringen*] itself. But arising
> means belonging to be-ing. (GA 65: 500–501, my translation; cf. CP 352)

> There is some evidence that language's ownmost emerging refuses to express
> itself in words—in the language, that is, in which we make statements about
> language. . . . [L]anguage holds back its own origin and so denies its emerg-
> ing as such to our usual notions. (GA 12: 175, my translation; cf. WL 81)

> Language's ownmost arising as such is saying as showing. (GA 12: 241, my
> translation; cf. WL 123)

> Saying must at first sound obscure and strange, yet it points to simple phe-
> nomena. (GA 12: 196/WL 101)

Notice, in the first place, that the question here is not about a defining
essence that would ground language conceptually and metaphysically. To

ask the question about what is ownmost to language in terms of arising and origin is already transformative, operating outside the enclosure of metaphysical thinking, which asks about being but not about the unthought of the first beginning: *arising*. Notice, too, the connection between the German words for origin (*Ursprung*) and arising (*Entspringen*) here. They evoke a dynamic sense of springing forth that resonates well with Heidegger's having said that words are like wellsprings that must be opened up and carefully tended and allowed to let what they say flow forth.

Since this thinking moves with the thinking of the first and other beginning (attempting to think be-ing), it would be good to remind ourselves that this thinking is attuned by *reservedness*. Heidegger here tells us that language holds back its own origin, refusing its emergence into our usual, metaphysical, defining words. Yes, something comes to language here, for thinking; but in this opening and broadening of the space for thinking both revealing and concealing are in play. As we shall see, it is this very play of revealing and concealing as opening and sheltering that (in one way of saying the matter) makes way for the arising and saying that we are attempting to think.

To attempt to let language say its arising and holding-sway as such, without attempting to fix it in some statement about language, calls for a particular kind of attentiveness to the movement of thinking that is carried by the words. This begins with the step back to where we belong and already are as speakers of language. We are not attempting to think something alien, something of which we have no experience whatsoever. We live in the midst of language, and we can let this resonate with what Heidegger says and think along with him and, if all goes well, think after him in our own ways, with the awareness of what opens up here. To leap directly into the heart of the matter, saying is showing. How simple that sounds in this context! Language arises and holds sway as saying. Saying in turn discloses itself as showing. Therefore, saying (*sagen*) is not simply identical with uttering words and speaking (*sprechen*) (GA 12: 246, 241/WL 126, 122). In both German and English this distinction is not hard to discover, even in everyday usage. We speak or talk but hardly ever simply "say." We always say *something*. There is no saying without content, without import, without something being shown. I may speak many words but have nothing to say: the words carry nothing of import; they show or disclose nothing.

If the matter were left here, it could seem that while saying means more than simply speaking it is, nevertheless, bound to speaking in that it seems to be necessarily a human activity. That is, it could be a particularly meaningful kind of speaking. However, that would be to ignore nonhuman saying. When someone says, "That painting really speaks to me," she is not just making use of an anthropomorphic metaphor. Even less was Thoreau doing so when he said that his gardening was "making the earth say beans instead of grass."[6] Therefore, Heidegger says, "saying is in no way the linguistic expression added to the phenomena after they have appeared. . . . We dare not consider showing as exclusively, or even decisively, the property of human activity. Self-showing appearance is the mark of the presence and absence of everything that comes into presence, of every kind and rank" (GA 65: 246, 242/WL 126, 123). Already in *Being and Time* disclosure was highlighted as central to understanding the meaning of our being. The t/here of Dasein is the opening for disclosure, and Dasein *is* its disclosedness (GA 2: 177/BT 171). Disclosure of *what*? Of its being-in-the-world, which means its understanding of the significance of the beings in its world. Quite obviously, those beings have somehow already shown up; they are somehow showing forth. They have something to say. So Heidegger there, foreshadowing what he does in *Contributions* and *On the Way to Language*, distinguishes between *Rede* (discursive disclosure) and language, with language arising from disclosure.

Saying and be-ing belong to each other. Saying shows forth from be-ing, and be-ing only comes to awareness as saying (which does not exclude in silence). *Saying* is another guideword in the thinking of being, in conjoining with all the others. In *On the Way to Language* and *Contributions* there are two more guidewords that need to be brought forward to open up what is coming to light: enowning (*Ereignis*) and way-making movement (*Bewëgung*). They are needed in order to head off the mistaken notion that saying simply articulates be-ing *directly* in language or even in perception. They need to be brought together with what we already hear here: saying (showing) marks *everything*. This "everything" will soon come more fully and explicitly into question. It will become clear why that is necessary in this preliminary discussion of enowning. It is preliminary to a much fuller discussion in chapter 3, where enowning figures largely in the timing-spacing of things (thinging). Here, in this discussion of saying, we are very close to making that region of questioning and thinking necessary.

"*The moving force in Showing of Saying is Owning*" (GA 12: 246/WL 127, Heidegger's emphasis). To open up the published translation a bit, it says: "The deep energy of the showing that moves saying is enowning." I have been using this word already without calling special attention to it. In many contexts in Heidegger the meaning is fairly clear; it suggests, in connection with things that arise and appear, that they are arising *into their own,* into what is uniquely their own way of arising and enduring as this or that thing. The way Heidegger uses the word "enowning" also intimates that this is not some isolated event; things are being-enowned, though this is not to be taken as some transitive, active-passive relationship between two beings, because the relationality here is much more complex than that. Once again, we are seeing the need to open up the thinking of the timing-spacing of things, soon to come. For now, we can note that bringing enowning into such a close joining with saying (showing) tells us something significant about *how* showing takes place. In ordinary German *Ereignis* means "event" or "happening." This has led some translators to link that sense to the *eigen* (own) that can be read into the word and translate it as "event of appropriation," which is not only incredibly stiff (canceling the dynamic energy of enowning) but also downright misleading. Heidegger is quite clear that "enowning, seen as it is shown by saying, cannot be represented as an occurrence or a happening" (GA 12: 247, my translation; cf. WL 27), much less as an "appropriation," with its overtones of subject-object grasping. What, then, does Heidegger want us to hear in this word *Ereignis,* enowning?

The connection of owning and showing carried by *Ereignis* is made more evident when Heidegger draws on an etymological trace to say, "Ereignis ist eignende Eräugnis" (GA 12: 253). We could read this as "enowning is the owning bringing-before-the-eye." *Eräugnis* (from *Auge,* eye) is a bringing-before-the-eye, bringing forth into disclosive appearance. Something *shows* up, just as it is, in its own way. This is not necessarily just to be taken literally, visually. Enowning as the heart of saying-showing may come into thinking as the arising of an insight, a decisive gathering up of the thinking experience around some matter, especially an insight into enowning itself, by whatever word it is invoked. It can also involve sound, silence, and dwelling with things (another preview, this time pointing forward to chapter 4).

The showing of enowning always remains in dynamic tension with what is not shown, in relation to both the showing of things and of language. In connection with language and thinking this has come up several times already as the play of revealing and concealing. That is also relevant to the later exploration of the timing-spacing of things. Neither language nor things are "boundlessly unconcealed"; there is always more than meets the eye or something held back. The *arising as such* of language and things always withdraws from thought and perception. *It shows itself only in the intimations of this withdrawing.* We can take that a little deeper, at least in a preliminary way. "*Ereignis* withdraws what is fully its own from bound-less unconcealment. Thought in terms of *Ereignis*, this means: in that sense it expropriates [*enteignet*] itself of itself. Expropriation [*Enteignis*] belongs to *Ereignis* as such. By this expropriation, *Ereignis* does not abandon itself—rather, it preserves what is its own" (TB 22–23). (Here I follow the pub-lished translation, except that I replace "appropriation" with *Ereignis* rather than the better translation of "enowning" so that we can move gradually through this difficult and subtle thought.) *Enteignis* is fairly nearly untrans-latable. "Expropriation" for it is no better than "appropriation" for *Ereignis*, but it does tell us one thing: *Enteignis* pulls in the other direction from *Ereignis*. To enowning, dynamically bringing everything into what is its own, belongs movement that goes the other way, too. How? In a different con-text the translators of *Contributions* translate *Enteignis* as dis-enowning (see GA 65: 231/CP 164). What does this mean? I think we can take a clue from something else in the guidewords that I listed in opening this section of the chapter. Heidegger says that what is ownmost to language, language's own arising and holding-sway, refuses to come to conceptual, propositional language and that this refusal or withholding belongs to its very arising as such, which denies its ownmost holding-sway, its emerging as such, *to our usual notions* (GA 12, 175/WL 81). What are our usual notions? Being. Presence. Essence. Subject. Object. Instead of "being" coming to word in a concept, we are here trying to think be-ing, which "is" enowning, which moves as saying-showing. "Being" is only an idea. "Be-ing" and "enown-ing" and "saying" are words that try to say something that is neither a being nor an idea. Any attempt to grasp and fix enowning will run up against its dis-enowning. And since enowning is not something that can be lifted out by itself but is always the enowning of things, it is, in a sense, dis-enowned of "itself." There is nothing here that can be fixed and reified. Enowning is ongoing dynamic relationality, which necessarily brings con-tinuous change. This again points forward to the discussion of timing-spacing in chapter 3.

This thinking moves in ab-ground, attuned by reservedness, which echoes the withdrawal of the matter for thinking from all of "our usual notions." Once again, we are reminded of the directives in "Memorial Address." To persist in the thinking of the first and other beginning we need to release the old philosophical presuppositions and remain open to mystery. Lan-guage's ownmost holding-sway (and the language of the holding-sway of enowning "itself") cannot be represented as can one of the beings of meta-physics. Yet, somehow, what is ownmost to language does come to language.

It comes to word in the language that *says* something, both disclosing and hiding its saying-showing movement. We are always already within language, never able to step outside it, distancing ourselves so as to objectify and grasp its ownmost arising and enduring as saying. It is both deeply mysterious and the nearest of the near. And it does open way(s), in play with our persistence in questioning, releasement toward things, and openness to mystery.

There are two neglected passages in *On the Way to Language* that reveal something more of saying's way making. (They are, in fact, so oddly underestimated and misunderstood that a significant part of one of them is dismissed as a "gloss" and left untranslated. See GA 12: 187/WL 91–92.) Just as saying is said to mark everything that arises, so too does way-making movement, *Be-wëgung*. This is not a standard German word; it is not just "movement" (*Bewegung*). The unusual hyphen and umlaut tell us to be attentive for something different, something more. Heidegger, exploring old links and possibilities in the history of the word, uncovers an old verb (*wëgen*) that means "to form a way, and forming it, to keep it ready. Way-making understood in this sense no longer means to move something up or down a path that is already there. It means to bring the way forth first of all, and thus to *be* the way" (GA 12: 249/WL 130). In the remainder of the two passages the manner of moving that makes way is described through bringing the word into play with some cognates that carry the senses of weighing, wagering, and waving (all of this plays forth in English as well). The thought unfolding here is that whether we are picturing scales, wagering (putting something in play, in the balance in an uncertain situation), waves rippling on a body of water, or a hand waving, there is a sense of motion that is decidedly not linear. It is more like resonating, vibrating, swinging, staying somehow in balance within some sort of back-and-forth or up-and-down motion. "[W]e are moving within language, which means moving on shifting ground, or still better, on the billowing waves of an ocean" (WHD 169/WCT 192). This evokes in powerful imagery the thought that the way for thinking (and thinging, as we will see shortly, in the next chapter) emerges in the unresolved playing forth of the dynamic of revealing and concealing, emerging and withdrawing. There is no way already *there* somewhere just waiting to be discovered and followed. Way-making movement gives and makes ways *in* way making. The way becomes way only as it opens and is thought or followed, in the same moving, at the same time. The hyphen in *Be-wëgung* and *be-wëgen* puts special emphasis on the prefix, which often turns transitive verbs into intransitive verbs. By emphasizing the prefix Heidegger may be suggesting that we are not to understand *be-wëgen* as a transitive verb in some typical subject-object structure. Way is not some object. Way making makes way in such a way that "it is" the way, that is, all there "is" is way-making movement. The movement moves, and that is all. It gives way in self-withdrawing, in yielding way. Such giving way clears the way for saying, for the self-showing of whatever is freed into the clearing or opening of the way. "But the pathway of this enthinking of be-ing does not yet have the firm line on the map. The territory first comes to be *through the pathway* and is unknown and unreckonable at every stage of the way. The more genuinely the way . . . is the way to be-ing, the more unconditionally it is attuned to and determined by be-ing itself" (GA 65: 86/CP 60). Attuning: reservedness. We know that well by now, and it is reinforced by what is said here.

However, we can get a bit more guidance in how such attuning works by way of something else that comes up in the discussion of the arising of saying (showing), the saying of things, and the saying that emerges into language. My question here is, Can we understand more clearly how say-ing emerges into our thinking and language? I have already pointed out that we are always already within language. We are also quite obviously always in the midst of things as well. But this isn't just a matter of location and history, of an accident of abstract time and space. "We hear saying only because we belong within it" (GA 12: 244/WL 124). In the German this comment is strengthened due to the play between *hören* (to hear) and *gehören* (to belong). Belonging within the play of saying-showing, in its display and its reticence, we can hear or attend to what shows forth and what only gives hints and intimations in its withdrawing. Belonging to say-ing places us within the way making of enowning, within "world-moving saying, the relation of all relations" (GA 12: 203/WL 107). Thinking then becomes our saying-after, or co-responding (*entsprechen*), to that within which we find our own arising in the dynamic relationality that can in no way be reduced to relationships between beings or between concepts (GA 12: 143/WL 52).

Wait a minute! Didn't Heidegger say that although this would sound strange at first "it points to simple phenomena"? Yes, he did. And he said this in another way as well, with a hint about why we find it so difficult to think anyway. "Situations are expressed in what was said which we find difficult of access for no other reason than their simplicity. At bottom, a specific access is not even needed here, because what must be thought about is somehow close to us in spite of everything. It is just that it is still hidden from our sight by those old-accustomed preconceptions which are so stubborn because they have their own truth" (WHD 98/WCT 152). This is more than just a reminder—yet again—about the importance of releasing and letting go of the old preconceptions. It hints also in saying that they *have their own truth,* that there was good reason why the Greeks stood in astonished wonder at the being of beings. Things, beings, are there all around us, not as some generic abstraction but one by one by one, in all their uniqueness. And they are, says Heidegger, arising in their own way within the same dynamic of be-ing and showing in which we find ourselves. If we *think,* then our language can co-respond to and resonate with saying, whether it is with the saying of the thinking of the first and other beginning or the saying of the Carolina wren and the daylily in the garden. But we are more used to "think of thinking" in connection with philosophy, not with trees, dogs, frogs, or stones. And now, in the thinking of the first and other beginning, being comes to language as be-ing. Be-ing is not a ground for beings but rather ab-ground. This rather strongly suggests that "beings" is not a concept that will hold up as we attempt the leap into ab-ground; it has lost its grounding. But then what of the tree, dog, rock, and wren? It is not as if they vanish. *There they are,* just out the window. But now, what are they? This is not the old question about the being of beings. This is a facet of the grounding question of the other beginning: what is the truth of be-ing? And what about what we still call "beings"? What "are" they now? Here, the old question "Why is there something rather than nothing?" is also a way to ask the grounding question (GA 65: 421/CP 297). This question, heard rightly, no longer asks for a ground (a "what," a being, or an essentializing definition) but asks a "why" (which is not so readily to be reified). We wonder how we can think these nearby beings now that "beings" has become a questionable matter. In Heidegger this question opens onto and circles back to what he first called "the horizon of being" in *Being and Time:* time. So next we need to bring together the thinking of be-ing, our question about "beings," and the matter of time.

3

Timing-Spacing-Thinging

One must be equipped for the inexhaustibility of the simple so that it no longer withdraws from him . . . [but can] be found again in each being. . . . But we attain the simple only by preserving each thing, each being—in the free-play of its mystery, and do not believe that we can seize be-ing by analyzing our already-firm knowledge of a thing's properties. (GA 65: 278–79/CP 196)

Their parametrical character obstructs the nature of time and space. Above all it conceals the relation of that nature to the nature of nearness. Simple as these relations are, they remain wholly inaccessible to calculative thinking. (GA 12: 201/WL 105)

With these thoughts Heidegger both gathers up some of the core insights of the first two chapters of this book and hints forward toward our attempt to take up what comes after (and departs from) the abrupt ending of *Being and Time*. There, Heidegger pointed to the need to *think time* as the horizon for the question of the meaning of being while (necessarily) leaving the task undone. Having situated *Being and Time* as thinking in the crossing of the first and other beginning and having explored what Heidegger means when he calls on us to *think*—along with the ways in which language makes way for such thinking—enables us to attempt to think with Heidegger as he returns to that matter. In accord with what has been said of language and thinking Heidegger here reminds us not to look for methods of approach to "time," or a theory of time, or assertions about time, or "new and improved" conceptual representations of time. Well, what, then, are we to seek?

If we are to understand how it is that this matter—the nature of time and space—can be wholly inaccessible to calculative thinking and yet be *simple*, we will indeed need to carry forward everything that has come out in the first two chapters. There have been many details, but they all tend to gather around these two core insights: (1) there is nothing here that can be reified, and (2) this thinking is carried not by standard linear reasoning but in the resonating interplay of what I am calling *joinings*. Regarding the first insight, the grounding function of being has been thoroughly shaken. "Being" is a concept and no more. And as for be-ing, "[b]e-ing itself is nothing in itself and nothing for a subject" (GA 65: 484/CP 341). The anything-but-nihilistic nature of this "nothing" will be noted near the end of this chapter and be a major topic of chapter 5. For now, just knowing that neither "being" (grounding presence) nor be-ing (ab-ground, neither present nor absent) can serve to ground our understanding of beings in the usual way is enough to *call into question* whether "beings" themselves are what we have thought they are. We cannot just presuppose or take as given our old notions about beings. Neither can we assume that our accustomed ways of acquiring and grasping knowledge about beings (method, technique, concept, representation, theory, system) are adequate to think be-ing *or beings*. We must let go of our attachment to those comforting devices and be open to the mystery that arises as we attempt to hold be-ing and beings in question.

Inescapably, our relation to language rises up at every turn. Written language necessarily takes a linear form, word by word, sentence by sentence, page by page; on and on it goes. However, the thinking of the first and other beginning is evoked and carried by joinings of texts, guidewords, questions, and more. The ways that joinings work to open up ways for thinking is only rarely linear. How they work is even difficult to say directly without resorting to hinting imagery. Joinings resonate, shimmer, echo, and bounce against and off each other. Like the jewels of Indra's net they mirror each other dynamically in such a way that the reflections, though belonging to each other, are never identical. Heidegger at times speaks of the movement of thinking as a "step back," but at other times he calls it a "leap" forward into the heart of the matter. We "step back," for instance, to retrieve the thinking of history of Western philosophy; in that instance we can see that such a "step" might be quite long-drawn-out. Or, pursuing another question, we "step back" to where we already are (e.g., as speakers of language) to understand it with new clarity. At times, "step back" and "leap" are nearly inseparable, just as in preparing to raise the question of the meaning of being, or preparing to attempt to think be-ing, we are already moving within the thinking of the first and other beginning. In that case preparing is already thinking; the thought of be-ing is not a result that is only to be obtained as the final result of some chain of thinking (WHD 48/WCT 12; GA 65: 227, 239/CP 161, 169).

One insight that emerges from *both* the nonreifiability of the words that carry the thinking of be-ing and the nonlinearity of that thinking is that the guidewords will have multiple meanings. Also, there will of necessity be more than one way into the heart of the matter, this *simple onefold*. Here is that thought again: in its very manifoldness, multiplicity, and complexity the matter itself is simple. It is simple in a quite literal sense of the word in that it is the one matter for thinking: be-ing. What I have mostly been calling "be-ing" actually can (and will and must) have many names, many paths and ways. This way of understanding the simplicity of the matter is not (though it might seem so at this point) a linguistic dodge. For Heidegger simplicity is the *gathering* of the onefold; it is what is the *same*, thought as what dynamically belongs together in its very differing. Granted, this is a very complex sort of simplicity, and to understand it more clearly or deeply is not easy. Even Heidegger had to struggle, as he tells us quite frankly. "On this 'way'—if stumbling and climbing can be called that—the same question of the 'meaning of being' is always asked, and *only* this question. And therefore the sites for questioning are constantly different. Each time that it asks more originally, every essential questioning must transform itself from the ground up. Here there is no gradual development" (GA 65: 84/CP 58–59). For now I suggest that we hold this simplicity of the matter in front of us as a question rather than an assertion and let it emerge along with the thinking. We can also take what Heidegger says about the simple, just above, as giving a strong hint about the way to focus our attention so as to understand the simple onefold without drifting off into a web of abstractions. This ever-so-complex simplicity comes to meet us in *each thing* we encounter around us.

I take that as a hint in helping us find our way into the question concerning the nature of time and space from among the many ways into the matter. How should we proceed? One way to engage the question concerning the arising and holding-sway of time would be to leap directly into sections 238–42 of *Contributions,* where we can find this topic taken up at some length. I think, though, that—given the unusual difficulty of the language there—taking that route would run the risk of our just following *words* and leaving the thought much too abstract. What makes Heidegger's thinking so powerfully transformative is that it is not just a set of conceptual abstractions. So let us take the hint I just noted: be-ing comes to meet us in *things.* Heidegger was, as a thinker, also a fine teacher, and if the students, hearers, and readers were really listening, they were led step by step *through* the words, through the abstractions, to the deeper movement of saying. But again, how linear this written language is! The "steps" (whether forward, back, or both at once) are not sharp-edged steps marching directly and evenly upward. They are steps more akin to those you see in old buildings on Greek islands: solidly joined, flowing together, with smooth, rounded edges, curving up or down, around and out of sight, perhaps with a brilliant red tomato sitting there quietly in the sun and the sound and smell of the sea all around. With that in mind we can now take up that hint and turn our thoughtful attention to what is all around us: things. Heidegger says that be-ing never reveals itself directly but comes to meet us in things. So I inquire into them simply as things, with the clear intention of attempting to let go of metaphysical presuppositions that could avert or sidetrack the inquiry.

THINGING

If we look around us, what do we find? Things. Books, zinnias, tomatoes, pen and paper, microwave oven. A thunderstorm is blowing in from the west. We would usually call a raindrop a thing. What about the rain? Probably. The wind? Perhaps. What about the thunderstorm itself—is it a thing? That is a bit harder to say for sure. What about the dog at the door, asking to be let in out of the rain? Is she a thing? Maybe, maybe not, depending on who you ask. But why do we answer these questions the way we do? On what are we basing these assumptions or assertions? Things are so close to us, so common, that we hardly ever ask, What is a thing? The word "thing" does not carry, at least at first glance, nearly the philosophical baggage that "being" does. But there are obviously some presuppositions about the nature of things that are at work and that play out in our thinking, our language, and our actions. By now we know that we are not looking for a definition, concept, or theory when we ask, What is a thing? but we can start there, as Heidegger did, to open a way into the matter.

We have inherited at least three influential concepts of the thing from metaphysics and epistemology. The subject-predicate structure of our language has been described as indicating *substances with their accidents*. The ball (the bare substance) is blue (an accident—it could just as well be red or yellow). Second, the thing has been defined as the *unity of* a manifold of *what is given to the senses*; the Kantian synthesis of perceptions into an object is one example. A hallmark of this concept of the thing is that it is inextricably bound to some theory of subjectivity. A thing is an object given to or synthesized by a subject. The third important concept of a thing is the hylomorphic idea: a thing is *formed matter*. Heidegger is quite definite that none of these notions of "thing" will serve. To think the thing in these conceptual terms is to miss our relationship with things, which is much more intimate than any of this theorizing allows. Things are not simply over-against us (as ob-jects). Substance and matter or pure, raw sense data are never what we actually relate to in our dealing with things. Just try it. Pick a thing, any thing. Where is the substance, apart from its so-called accidents? Find me a raw sense datum. Of course, we can infer that such things *might* be there, somewhere and somehow. But I am not asking about any possible inferable notion; I am asking about a thing and how it actually shows itself to us or what it has to say to us. Here is a thing, not far away at all.

A loaf of plain, homemade bread stands hot and fragrant on the cutting board. I am not going to leap to the question, What is it, this thing, this loaf? Our philosophical tradition shapes our day-to-day thinking to the extent that asking a question worded in that way is to ask for a definition or essence. But that is precisely what we do not want just now, a conceptual definition on which we can rest and set aside the question. I ask instead, What gives us this loaf? I am not yet asking for some ultimate source or ground, which, once again, would tend to lead us to a metaphysical answer, closing off further inquiry. I just want, to begin with, a down-to-earth description of what goes into making up and giving us this loaf, right here in front of us, soon to be cut, spread with blackberry-elderberry jelly, and eaten. We can begin with the recipe. This happens to be a loaf of semolina-sesame bread. So we have wheat flour (unbleached bread flour and high-protein golden durum flour), water, olive oil, salt, sesame seeds, and yeast. The flour, oil, and seeds were grown somewhere, so we also have what was needed in the growing: the soil in which the crops grew, the sun shining on them, the rain for their germination and growth, and the farmers to plant, tend, and harvest. But the grain and seeds in the field and the olives in the grove and later in the store are not yet the loaf of bread. They have to come together with the water and the yeast. So we need a baker (me, in this case) and some equipment: a bowl, a sturdy olivewood mixing spoon, measuring cups and spoons, a counter on which to knead the bread, a pan, and an oven. Potholders, cooling rack, bread knife. And then, of course, someone is drawn into the kitchen by the smell of the baking to eat the bread hot out of the oven and in tomorrow's sandwiches. That's just the loaf of bread. We could ask in the same way about the bowl, the spoon, the oven, the jelly, and so on. For each of these there will be a web of things that were necessary for that particular thing to come into its own and arrive in this kitchen to come together as this loaf of bread. If we seriously try to trace out these webs of necessarily related things inclusively, it is hard to see where they could end. The sunlight to grow the grain only comes to us in just the right way to grow crops because our planet happens to be at just this distance from the sun, about 93 million miles. But that, in turn, depends on the gravitational relations at work in the entire solar system. If Jupiter were significantly smaller, we would effectively be *in* the oven, never mind having the chance to bake bread. So what do we have here? Just a little careful thinking in response to a question about one particular thing shows that many other varied kinds of things and activities and processes must be there in order for that thing to be here. This is the same simple observation that Heidegger makes in his considering a pair of painted peasant shoes in "The Origin of the Work of Art" and a jug in "The Thing" (GA 5: 1–74/PLT 17–87; GA 7: 167–87/PLT 165–86).

Having done that we can now ask, What is a thing? without looking for a metaphysical or epistemological concept or essentializing definition of the thing. What is ownmost to a thing as thing? How does a thing arise and endure and hold sway as such? Heidegger: "The thing things. Thinging gathers. Bringing the fourfold into its own [*ereignend*], it gathers the fourfold's stay, its while, into something that stays for a while: into this thing, that thing" (GA 7: 175/PLT 174). The thing things. Not only is the thing intrinsically relational, it is *dynamic*. To think of a thing as inert substance, or matter, or object is to reduce it to much less than what it is. The thing is what it does: thinging. But what is thinging? Thinging is *gathering*. This is not a gathering up or synthesizing of sense perception, which is the activity of some subject. The *thing* things, that is, gathers. It gathers together as a thing. What is gathered in this gathering that is the thinging of the thing? "This . . . simple onefold of sky and earth, mortals and divinities . . . [that] we call the world" (GA 7: 181/PLT 179). The world named here is not some abstract metaphysical whole; it is not the sum of beings or even of things. Rather, it says how we humans meaningfully experience or journey through the dynamic clearing in which we find ourselves in the midst of things. The world: earth and sky, divinities and mortals. Earth: solid, living, bearing up in stone and sea, bearing forth in grass and flesh. Sky: sun and stars, day and night, the cycles that surround us. Mortals: we humans, capable of knowing death as death. Divinities: the beckoning intimation of what continually withdraws from us, the presenc-ing, thing-ing, gather-ing movement itself, hinted at in things that evoke a sense of awe or reverence. The "divinities" here do not name the movement itself but its intimation of mystery.[1] This thing-ing movement, the gathering of the fourfold, is itself no-thing, yet without it: no things. Earth, sky, mortals, and divinities gather and disclose themselves as such in and only in the thinging of the thing and its echoes in language. These are not four metaphysical entities or categories (metaphysical or epistemological) prior to and apart from thinging. They are nothing that can be thought as transcendent to things. The fourfold is not meant as a categorical outline of "what there is." It is a word that says and thus shows something of what takes place in the thinging of the thing. It could be a threefold or a fivefold. What is *said* is that all things are involved in each thing. There are no isolated, independent, substantially self-existing things. *Things gather the fourfold in and as themselves* (GA 7: 40, 150/PLT 53, 149–50).

Heidegger uses another kind of imagery to attempt to say this thinging of the thing. "The mirror-play of world is the round dance of [enowning]. . . . The round dance is the ring that joins while it plays as mirroring. . . . Out of the ringing mirror-play the thinging of the thing takes place" (GA 7: 181/PLT 180). Dancing: mirroring: playing: ringing. Playful circling joins and parts and entwines the fourfold's dance of emerging and withdrawing, showing and hiding, in an arranging and rearranging without ultimate cause or goal. Ringing is the dancing and the resounding and the contending, where the echoing and resonating of things with one another is not mere accord but also a wrestling of one another into an opening where they may show themselves. In the round dance of enowning gathering and dif-fering (literally, carrying apart) strive and play. Each dances unto its own in the owning of each to the other, in the enowning that the fourfold enacts and says. In mirror-play there is no mirror (being, ground, god), only the dynamic mirroring. This ringing and resounding is utterly nonrepresentational, loosening the mirroring even more from the notion of mirror image to a mirror-ing that is a clearing of the "between" for thinging, the gathering of the fourfold in their entwining resonance, their mutual enfolding-unfolding. It is the ringing of stillness we thought with Heidegger in chapter 2.

Return to the loaf of semolina-sesame bread standing there just out of the oven. It gathers the fourfold of earth (the soil in which the wheat grew, the wheat itself, the yeast, the water, the pottery bowl, and the wooden spoon), sky (the sun shining on the growing, ripening wheat, the rain for germination and growth), mortals (the farmer, the baker, the eater), and divinities (the hint of the mystery of the bringing near of earth, sky, and mortals, of the ever self-concealing movement of thinging). The loaf of bread is all these things; they make the loaf what it is, bringing it into its own. Yet the loaf of bread is not simply all these things; it is not some unity or combination of the fourfold. The loaf of bread is a loaf of bread, gathered to itself, showing itself as something dif-ferent, something carried apart, from every other loaf of bread and from all other things. In the very gathering of the fourfold, the bringing near that is thinging, this differing occurs and continues to occur, for no thing endures unchanged. A simple loaf of bread is an ever-changing web of fluid and complex relations. Toast. Sandwich. Bread pudding. Crumbs for the birds. Things are radically changeable, finite and multiple, and empty of independent or substantial self-nature. These things gather and unfold, disclose themselves, show or say to us some of what is their own. This saying-showing is the same saying that is the heart's core of language.

All this emphasis on movement certainly acts to counter even more the old, already-shaken notion of being as presence and of what is most "in being" as what is continually present and unchanging. But we might be tempted—conditioned as we are to seek answers, to look for something on which to rest and say, "That's it!"—to take that to an extreme and turn this into some sort of "process philosophy," with movement as the definitive ground for explanation. Certainly, it is this very movement that makes way for things and their saying-showing and gives to us the possibility of being t/here and saying something in response to this showing forth. But "movement" is not some ultimate category. "Saying, as way-making movement of the world's fourfold, gathers all things up into the nearness of face-to-face encounter, and does so as quietly as time times, as space spaces, as quietly as the play of time-space is enacted. The soundless gathering call . . . we call the ringing of stillness," the language of arising as such (*Wesen*) (GA 12: 203–4/WL 108). This thought collects what we have so far and spreads it out into several significant paths for further thinking. This "ringing of stillness" is, in the obvious sense of what is said here, the silence of the gathering of things into their own. There is no "and God said, 'Let there be X'" to be found here. But it should also be fairly obvious that there is much more to this thought. Whether they are the things of nature or of manufacture or of agriculture and gardening, "thinging" is quite often rather noisy! So the literal sense, while not exactly incorrect, is rather superficial. The ringing of stillness is not something other than the ringing mirror-play of thinging. What we have here is a joining that again gathers many facets of the one matter. The stillness that rings forth intimates the withdrawing of the ringing itself, arising itself, gathering itself. *Ringing* carries not only the *dancing* sense that Heidegger draws from the history of the word but also *sounding*: echoing, resonating, and reverberating, all of which carry a sense of nonlinear motion. In this moving there is also stilling, stillness, whereby things come into their own in a way that carries pattern and sense (the French *sens* is helpful here, conveying the nuance of direction as well as meaning). This is, after all, the dynamic stilling of *saying*, which says (shows) the things but only hints at much that is gathered in them as well as the gathering itself.

This guiding thought also says that gathering involves bringing things near in the play of time-space. As nearing, gathering brings things near to one another. This is not so much meant spatially (in terms of our usual notions of parametric or measurable space) but more in the sense in which we say of people, "They are very close to one another." Part of what they are (what is their own) is their relationship with the other person. More: part of what is their own *is* the other, and vice versa. Yet in this bringing near nearing preserves farness. Nearing, bringing near, preserves what is each thing's *own*. It is not a blending or joining into an undifferentiated oneness. Nearing is also a differencing, dif-fering (carrying apart and carrying out) each thing from the other. The gathering that is thinging is a mirror-play of nearing and distancing in which each thing comes into its own. The distancing and nearing are the same: they *belong* together (GA 7: 179/PLT 177–78). To understand this better, let's bring it a bit more down to earth. This will also move us closer to being able to bring time and space into the discussion.

Gathering is relationally dynamic. And in the relationality the gathering is mutual. The bread bowl is what it is, not only in its gathering of clay and potter and fire but *also* in its gathering of flour and water and yeast and the actions of the baker in this very kitchen (with all the larger and larger web of relations that are implicated as we add in each facet). Without bowl, no bread. Without bread, no bowl. We can say that "in general." But also: without *this* bowl (its weight, its density, how it retains warmth "just so" for the rising) there would not be *this* loaf. And without this loaf, and all the other loaves and salads and mashed potatoes made in this kitchen with these hands, it would not be this bowl, here and now. I am not talking about creation or causation but about mutual or interdependent arising. Likewise, after the sketch of the fourfold gathering-thinging

in "Building Dwelling Thinking," Heidegger discusses a bridge as an example. "The banks emerge as banks only as the bridge crosses the stream. . . .
[T]he bridge brings to the stream the one and the other expanse of the
landscape. . . . It brings the stream and bank and land into each other's
neighborhood. . . . Always and ever differently the bridge escorts the lin-
gering and hastening ways of men to and fro" (GA 7: 154–55/PLT 152). This
bank and that bank (which already suggests that I—or someone—is standing on one side or the other) are "just so" due to the bridge joining them
for our crossing from one place to another, carrying out our varied activities. We are so used to hearing "thing" as some neutral, inert object that
to begin to think "thing" as dynamic, ever-changing gathering of a multifaceted relational web is already transformative. It also hints at another
question that will come up even more urgently (and more than once) later.
We do not usually call ourselves "things," but doesn't this gathering unfold
in and as us, too? If we read *Being and Time*'s account of Dasein's being-in-
the-world, it tells us that Dasein is being-there, which is to say, it is t/here
for the disclosing of the beings of its world. Heidegger put it even more
strongly, as I already noted in chapter 1: *Dasein is its disclosedness.* It is none
other than opening for disclosure, for—as we can now call it—saying-
showing. But Dasein—this opening—is us. I want to ask at this point, Is
there any compelling reason to think that our own arising is radically different from that of all things? In raising the question of the meaning of
being Heidegger said that we ourselves are put in question. Surely that
questioning will come forth in very specific ways as we press on in thinking. I will not attempt to answer my question now because it will come
up again and again until the "answer" comes forth with the question.

    The gathering of things is, Heidegger said, a gathering into nearness in
the play of time-space. The joining begins to enlarge as Indra's net grows.
The discussion of thinging in "Building Dwelling Thinking" focuses on
space as it carries forward with laying out what can be thought with keen
attentiveness to that bridge. The fourfold, the dynamic web of relations,
comes together in the location of the bridge, making a site for it. There is
no such site, however, until the bridge begins to emerge under the hands
of the construction workers and their tools. Follow along with Heidegger
as he carries the thought farther along. "Only things that are locations in
this manner allow for spaces. What the word for space . . . designates is
said by its ancient meaning. *Raum* [room, space] means a place cleared or
freed for settlement and lodging . . . within a boundary. . . . A boundary
is not that at which something stops but . . . that from which something

begins its presencing . . . the horizon. . . . *Accordingly, spaces receive their being from locations and not from 'space'"* (GA 7: 156/PLT 154). In short, gathering is thinging, which yields a location or site for the thing to come into its own, which only *then* can be seen as a space for its ongoing presencing. The very idea of "space," even in a fairly concrete sense, depends on things and their locations.

What, then, of our modern Cartesian-Newtonian notion of space as a featureless, unchanging, empty expanse that can be divided into a mathematical grid on which to measure the extent and mark the locations of things? Heidegger shows how this notion of space is at yet another level of abstraction, being derivable from the spaces that emerge in thinging. He keeps to the example of the bridge. The mutual thinging by which bridge and road and banks and their landscapes, with all their related features, mutually condition or be-thing each other (in German, the word for "condition" in this sense is *Be-ding*, which is quite literally "be-thing") *also* opens the locations, the places where each thing gathers. If, however, these places are—as they can be and have been—treated as mere positions, then the gap or interval between them can be measured, and "space" then becomes just another word for interval (distance). In fact, our English word "space" comes from the Latin *spatium,* which primarily means "distance," and—to show what a derivative latecomer this notion is—was not used to name the abstract mathematical expanse or to refer to things as "spatial" in that sense until the nineteenth century.[2] Heidegger's own very clear unfolding of the further derivation needs no elaboration of mine. "The bridge now appears as a mere something at some position, which can be occupied at any time by something else or replaced by a mere marker. What is more, the mere dimensions of height, breadth, and depth can be abstracted from space as intervals . . . we represent as the pure manifold of the three dimensions. Yet the room made by this manifold is also no longer determined by distances. . . . [It is] now no more than mere . . . extension. But from space as [extension] a further abstraction can be made, to analytic-algebraic relations. . . . The space provided for in this mathematical manner may be called 'space,' . . . [b]ut it contains no spaces and no places . . . [or] things" (GA 7: 157/PLT 155). "Space" is abstracted from "locations," which in turn depend on things already being there. Notice the flattening out of the dynamic relationality shown by things. By the time the increasing levels of abstraction finally arrive at "space" in the modern sense, things have been reduced to extended objects at such and such a position that take up a certain measurable volume.

It is not at all a stretch to see this as part of the long-developing trend that now, at its extreme, allows things (and even us) to be taken as interchangeable units of a standing reserve of things on call for use in production of various kinds. This is, of course, far from the only factor. The idea of abstract mathematical space emerged just about the time that rigidly conceived subject-object dualism took hold, as set out most famously by Descartes in *Meditations on First Philosophy* and *Discourse on Method*. He, in fact, defined "body" (which is as close as he could get to talking about things) as extended, utterly inert, dumb substance. Only mind (rarefied and unextended but somehow substantially existing) was able to act, to take charge and make things happen, or, for that matter, to perceive (which he subsumed under "thinking"). It is fairly easy to see how this idea was not pulled out of thin air (or out of "space") but rather built on and developed the thinking that unfolds from the first beginning of Western philosophy with the Greeks. The increasing abstraction, rarefying, and mathematization of space and time, from Aristotle's notion of time as the measure of motion, of the "count" of the now-points, down to our contemporary notion of the reality of ultraprecise clock time, playing out in the way our lives are dominated by public time, depends not only on early shifts in thinking about time and space but also on assumptions about thinking and about the kind of language that carries it. Thinking becomes more and more associated with *psyche*, with the intellect (instead of embodied sensory awareness), with the rational soul (which is, according to Aristotle, possessed only by human males), and with the soul posited in the monotheistic religions. These notions arise in decided and growing opposition to body (body arising in and as *physis*), which is variously seen as being entangled in and deluded by the sensory shadows of the forms, acting out the urges of the sensitive and appetitive souls (nonrational, more akin to the functions of plants and animals), something to be abandoned at death when the soul goes on to heaven, and finally—with Descartes—nothing more than a machine.[3] The layer-upon-layer development that intensifies this core dualism is entwined with several other value-laden dualities, including the assumption of the superiority of *logos* over *mythos* and the notion that the intelligible is closer to the real than is the sensible. In addition to the first beginning's invention of what we can call the ontological difference and the decisive character of the shift from oral culture to increasing reliance on the written text, we should recall what accompanies both: a decisive shift in focus away from nature (*physis* as the arising of things as well as its saying or showing) to exclusively human (and mental) conceptual, representational, linear, written language. The kind of thinking that is now valued most highly is that carried on with the least regard for body and for the web of meaningful, dynamic relations in which it arises. There were a great many intertwined transformations that brought us to where we are now. In thinking toward an other beginning we may expect many interrelated transformations to emerge as well.

Heidegger's account of the derivative nature of space, with its correlative flattening of thinging's dynamic gathering and arising, reminds us of his account of the flattening of temporality that takes place when the abstract, parametric notion of time becomes dominant, as discussed in chapter 1. In brief, Dasein's being, its there-ness, is described as being open for disclosure, with this opening structured by temporality. Temporality is not "time" but is the dynamic temporalizing of Dasein's being-*in-the-world*, in which a web of meaning, of significations, arises for Dasein to take up, ignore, or reject. Temporality is structured by (1) finding ourselves already in a certain situation (the ongoing unfolding of the things that have shaped us in our world), (2) our being now enmeshed in the midst of beings with their meanings, and (3) understanding and taking up possibilities to move toward, with all of this linked through (4) discursive disclosure. Heidegger emphasizes the nonlinear, dynamically simultaneous character of this temporalizing and the way that discursive disclosure (foreshadowing later discussions of saying-as-showing) pervades the entire structure, as Dasein then articulates the disclosing in language and thought. So the development of the idea of time as the linear sequence of now-points, each distinctly separate and even isolated from all the others, along with the correlative narrowing of arising and presenc*ing* to presence (being), *flattens* the experience (*Erfahrung*) of being-in-the-world (journeying through the play of temporality). It is reduced to the experiences (*Erlebnisse*) of bits and pieces strung along a line, grasping at whatever can be subsumed under the dominant modes of thought and the actions they ground, eventually becoming (for the most part) channeled into the narrow, one-track, techno-calculative mode.

David Abram relates another way that he uncovered the derivative and impoverished character of abstract, parametric time while opening onto a more originary, embodied experience of temporalizing. Going to a wide field or low hill, he closes his eyes and imagines his entire past, not as some measurable line of events but as the whole mass (here, I tend to picture a cloud, with no set geometric shape) of what has led him to this place and time, all just there behind him. In front is the mass of all possibilities yet to come (an even more amorphous cloud). The clouds or, as he calls them, "bulbs," joining together like an hourglass, meet in him at this moment. They begin, ever so slowly, to flow into the present, and as they shrink it grows until, opening his eyes, he says, "I find myself standing in an eternity, a vast and inexhaustible present" that is filled with all the sounds and sights and smells of the meadow, filled with life and movement: oaks and shrubs conversing on the breeze, ants and crows, rocks and lichens. "For my body is at home, in this open present, with its mind. And this is no mere illusion. . . . I am embedded in this open moment."[4] This "open moment" is the moment of the thinging of the thing. As *open*, this moment is in no way a now-point. Reflecting on this experience, Abram draws on (1) Maurice Merleau-Ponty's insight that we dynamically embody timespace (in the chiasmatic intertwining of what he calls "flesh," which is the world's every bit as much as ours) and (2) Heidegger's account of temporality to try to bring some account of what took place in the meadow to language. What can be said, beyond just giving a description of an experience? Merleau-Ponty, in his last, unfinished work, gives a clue that suggests that the "past" is simultaneously (and dynamically) *inside* this flesh.[5] What, then, of the future? Abram calls on Heidegger's hint that time is the horizon for thinking being along with the later comment in "Time and Being" that what is yet to come *withholds* its presencing, while what has been *refuses* to come to presence. Abram's thinking along with Heidegger here is impeccable. "While in *Being and Time* Heidegger wrote of the centrifugal, ecstatic character of time—of time as that which draws us outside of ourselves, opening us to what is other—in his later essay he stresses the centripetal, inward-extending nature of time . . . as a mystery that continually approaches us from beyond, extending and offering the gift of presence while nevertheless withdrawing."[6] So then, with Heidegger, Abram says that the refusal of what was and the withholding of what will or what may come are an "extending opening up" that gives all presencing.[7] Bringing these hints to bear on his experience in the open meadow, Abram asks where and how we experience that refusal and withholding that dynamically extend and open up for presencing and disclosing. What is *inside* (under the ground and in our—and others'—bodies), supporting perceiving, is refused to our direct perception. What is beyond the horizon (holding open the arena for perceivable presencing) is withheld. Both of these are unseen but very much in play in every immense moment. "And this living terrain is supported not only by that more settled or sedimented past under the ground, but by an immanent past inside each tree, within each blade of grass, within even the tissues and cells of our bodies."[8] This thought yields insights and questions that will continue to reverberate throughout my thinking along with and after Heidegger.

Neither Abram nor Heidegger intends what they say about the flattening of temporalizing and the derivative character of "time" to be taken in the direction of establishing a more correct concept of time (GA 65: 74/CP 51). What Abram does here reinforces Heidegger's insight that we do not actually experience time (or space) as autonomously existing dimensions (whether line or infinite void) that transcend us but as ways that our embodied experience is structured, dynamically holding open everything we can possibly sense or think. And just as Heidegger does in "Building Dwelling Thinking," Abram comes to the thought of "space" as a dynamically relational field in which we "rediscover the enveloping earth."[9] But, again like Heidegger, Abram is not proposing this as some kind of comprehensive theory of space. Taking up an issue relatively untouched by Heidegger, I share Abram's doubt that we are justified in this (and many another) area in assuming that *our* embodied experiencing is altogether and radically different from that of the other animals who embody a shared evolutionary past. The "within-ness" of what we call the past is much more a basis for acknowledging kinship than separation. And even in terms of what we call the future this may be so. We often hear the claim that animals have no sense of time. While it is true that I have never seen a cat consult a watch, we have all heard accounts of dogs meeting the school bus at the same time every day. Freddie the yellow Labrador embodies not only his genetically endowed instincts but also all of his past experiencing, and he enacts it as he moves toward his horizon, fully expecting little Jenny to emerge from that big yellow contraption that is not yet even in sight. Certainly, all the differing kinds of animals have their own unique capacities (Freddie can hear and without any doubt smell the bus coming long before any of us; we, on the other hand, can say that it will come at 3:35 and set an alarm to remind us to go meet it). But I suspect that what we share is much more than what sets us apart. Even in the matter of language (which our received philosophical and religious tradition takes as something decisively separating us from all other solidly embodied existence) there are, of course, various kinds of philosophical and scientific debates ongoing about whether or not the other animals "have language." They tend to define the "problem" in such a way that the answer to the question is most often a "no." This is yet another example of how method—whether scientific or philosophical—constrains not only the approach to a problem but even what counts as an answer. I find that calculative, problem-solving approach much less interesting than the implications of Heidegger's insight that *saying*, understood as meaningful showing, is a mark of *everything* that comes to presence, particularly if we are very careful in how we interpret that in relation to our fellow sentient beings (assuming too much is as un-helpful as assuming too little). Of course, insights from scientific research may well help us in thinking about a matter such as this. I will take this question up again, in more depth, in chapter 4.

One thing is quite apparent here: the sheer energy, the power of language and thought to bring things to presence, to—effectively—bring "a being" into existence where it had not been at all before, and to change not just how we talk and think but how we understand ourselves and live our lives. In thinking the first and other beginning we uncovered the thinking that, in hindsight, can be said to have created the ontological difference and, in so doing, created the possibility of reifying "being" understood as something in and of itself, to ground all other beings. Unfolding from that beginning, thinking conceives of parametric time as an infinite sequence of nows. As I said already, try finding one such now; no matter how you imagine divid-ing up a now (after *imagining* it in the first place), it will still just always be now. To say in response that I should know that a point is defined as having no extension does not adequately sidestep the question. What is this "now"? As far as I can tell, it is just that definition, it is "just" a con-cept, though one with nearly all-encompassing power in its playing forth in our contemporary world. Thus far I have focused somewhat more on how all of this has tended to flatten experience and constrict thought, but it is this very same creative energy of language—or, better, of the *saying* or showing that is the heart's core of language's arising—that also holds open the possibility of opening ways that could decisively change and broaden our thinking and (perhaps) free us and things from being simply trapped in enframing. This serves as another reminder to stay attentive to the need to release old assumptions and to remain open to mystery as we proceed.

I will return now to thinking along with Heidegger as he unfolds some of the ramifications of his account of thinging and the derivative charac-ter of parametric space in "Building Dwelling Thinking." We recall that spaces are yielded by the locations that gather as things emerge in dynamic relationality. What about us? How are we situated in regard to space? In the first place, we (the mortals) are part of the fourfold that is gathered in thinging. Furthermore, "space is not something that faces [us]. . . . It is neither an external object nor an inner experience. . . . We always go through spaces in such a way that we already experience them by staying constantly with near and remote locations and things. . . . I am never here only, as this encapsulated body; rather, I am there, that is, I already pervade the room, and only thus can I go through it" (GA 7: 158–59/PLT 156–57).[10] Neither we nor things are "encapsulated bodies." We are, one and all, shimmering interwoven dynamic relationality, through and through. And *all the way down*, into (within) ab-ground. This will become more clear after we take a close look at Heidegger's attempt to think time, space, and thinging together in *Contributions* and the later texts that move as joinings with it.

TIMING-SPACING

The extended account of time-space in the joining entitled "Grounding" in *Contributions to Philosophy* is, to say the least, difficult reading. The beginning of the subheading that contains these sections gives us an indication of why that is. Its title reads "Time-Space as Ab-ground," and the text opens with this question: "In what way of questioning is the so-named [time-space] embarked upon?" (GA 65: 371/CP 259). The question concerning time-space moves, as does the thinking of being, in(to) ab-ground or, to say it with more precision, into the thought of be-ing, in terms of time-space, *as* ab-ground. No matter how often we remind ourselves that be-ing is not a being, and that ab-ground is not something we could picture like a huge, gaping abyss, and that it is "only" the staying away of ground (something that was, in the first place, created conceptually), it is difficult to *stay with* this thinking without lapsing into grasping at representation. What are we attempting to think? (1) Be-ing that is not (a) being, though for centuries many assumed it was, (2) ab-ground, and now also (3) time-space that is not going to be grasped as "time" or "space." These are incredibly elusive thoughts, even more so in that they all *say the same*. Again, we have already accounted for that elusiveness to an extent in an earlier discussion of the way that what withdraws from before genuine thinking is what draws thinking along and keeps it moving. The elusiveness of the language echoes the elusive withdrawal of what it is attempting to think. However, that does not prevent this from at times seeming not just elusive but impenetrable. On the other hand, we have Heidegger's comments that all of these guidewords name one simple matter and that the relations here are actually quite simple. My task here is to move from the seeming impenetrability of some of the language toward some insight into what is simple in the saying, in what is actually showing itself to us, if we can think it.

The first step is to remind ourselves of the context in which this questioning account of time-space finds its place. The context that opens the way that has led us here is, of course, the thinking of the first and other beginning of Western philosophy, a way of thinking that even in its most preparatory stages is both a step back (retrieving the initiating thinking of the Greeks) and a leap forward into the possibility of an other beginning. In thinking the first beginning as beginning an other beginning is already intimated. The creative dynamic of the first beginning not only initiated the long and rich history of Western thought but also set in place ideas and trends so powerful that as they converge with correlative impulses toward security and control they eventually tend to limit thought and patterns of action to such an extent as to endanger the earth itself. The early wonder at beings' arising into presence, into the here (space in a simple sense) and now (time, not yet a mathematical now), is flattened into beings that are grounded on being and measurably located in vacuous space and eternal, inexorably ongoing, linear time. Dualism of various kinds further reifies the ideas that arise, more and more separating us from the capacity to sense our kinship with what the Greeks called *physis* and we often call nature. ("Nature" is, in fact, one way to translate *physis* as it refers to what arises of itself rather than being made by us or by some god.) The outcome is what we see all around us: the things of nature are seen as less than objects, along with many of us. If at times we find ourselves thinking, "This is insane!" I can only agree with Abram that to cut ourselves off from the sources of our capacity to speak and think is, in a sense, to lose our minds. Abram points out that our society's genocidal assaults on Native Americans and other tribal people in forcing them off their native soil cuts them off from their "matrix of discursive meaning" and thus drives them out of their minds. I am reminded of what the holy man Black Elk said in his first-person account of how his people, the Lakota, changed in undergoing such assaults. As the situation developed, both when the Lakota were being hounded to surrender and move to reservations (about 1870–90), and afterward, on the reservation under the control of the culturally genocidal Bureau of Indian Affairs, he gives examples of this "loss of mind." People were unable to hold to the old values and ways, and so they lashed out in desperation, at times treating both humans and animals with a brutality that would have previously been considered utterly bizarre and unacceptable.[11] Ironically, what our Euro-American society has done to others it is also—and has been for quite a while—doing to us. Only the degree of gross brutality differs. The sharper our calculative minds become, the less likely it is that our embodied intelligence will be able to actually *think* in any other way or to hold back from destroying the earth, our matrix of meaning, in our blind and compulsive desire to control everything.

Thinking with Heidegger this far, we are already engaged in a leap into the heart of the matter, the transformation of the grip of the concept of being, to the thought of be-ing. To see the first beginning with clarity *as* beginning and in its ways of unfolding is to see it in its optional character. It happened, but it could have happened otherwise. The thought of being might not have arisen in just that way. It was amazing and powerful, and some of the thinking that arose from it accomplished great things. Nevertheless, once we have thought this far we will never be limited in just the same way again. If there is an easy thing to see here, that is it: the optional character of the ideas that arose in the first beginning. The hard part is to take the thinking farther, faced with such openness and such difference from our usual ideas and ways of thinking. We are, again, looking at something akin to Indra's net, with multiple mirroring-joinings. One could take up any one jewel and think through the others. I have taken up the thing and thought its gathering to this point. Before going farther, I think it would be helpful to recapitulate some of the key points about the nature of thinking to see how they also lead into the question of the arising of time-space as well as helping us to think it.

In chapter 2 I began with suggesting that we attempt this thinking by holding in front of us two hints given by Heidegger. Releasement toward things and openness to mystery, he said, may enable us to start thinking. Releasement toward things, as I related it, begins with a willingness to let go of the "things" emerging from and determined by metaphysical and calculative thinking: method as setting means and ways, the theoretical results, and language understood only as grasping concepts, representations, definitions. In sum, we attempt to let go of clinging to rigid, linear, dualistic, goal-oriented ways of thinking because we can see their tendency to close down questioning or, at best, to narrow thought into one track. We first came to a sense of what openness to mystery might mean by way of considering the language that carries Heidegger's thinking, since it seems quite likely that that will also help us in our thinking after Heidegger. To encounter words, thoughts, and even texts as *joinings* that say or show something significant without rigidly fixing it or reifying it is my starting point. In a way, joinings take the place of two aspects of traditional thinking: (1) joinings *make ways* rather than set up a method, and (2) they do not yield concepts, theories, and so on. The key is to stay with their *dynamic* relationality, their mirroring play that refuses to fall to either extreme of identity or difference, instead "saying the same," gathering and carrying out (dif-fering) the complexly simple "one matter." But that is not all there is to say about joinings.

We just went through what Heidegger has to say about things, which came into the discussion here in part because that is where the idea of space comes in for the most attention as a distinct question. There, it is clear that for Heidegger "things" means *thinging*, which is, yet again, dynamic relationality. Some clarity about thinging is crucially significant in attempting to understand Heidegger's thinking of time-space, which is to say, to think be-ing as timing-spacing. Thinging is the open, dynamic gathering of the webs of relationships into each unique but always-already-changing thing that arises. Pause and think about that: it is remarkably like what I have said about the joinings of the language that carries thinking. In both of them, in the *joinings of thinging* and the *joinings of saying*, we have:

1. Gathering
2. of what differs but belongs and comes together,
3. whereby everything involved in the gathering is changed and continues to change
4. as they mutually mirror and intertwine with one another
5. and refuse to be grasped in any definitive, limiting concept.

The joinings of saying, manifesting as guidewords for thinking, echo the joinings of thinging. Just as is the case with an echo in a valley between two sinuous hillsides, the echo is not an exact duplicate of the sound that initiates it. And if the echo repeats and resonates multiple times, it is still recognizably still the same but more and more obviously not identical. It says the sound but dif-fers, carrying it out and apart in new relationships with the surroundings, until it fades into the ringing of stillness whence it arose. This is another way of thinking more deeply into what Heidegger meant when he said that saying or showing is a mark of *everything*, not just of human speaking and language (GA 12: 242/WL 123). If thinking is to resonate in accord with what shows forth, it then must echo or corespond with the saying that initiates it in thinging and in language. Many guidewords have emerged as jewels, as echoes, as hints, as way stations on the path of thinking. Some of these we will find helpful to remember as we move into the region of thinking (questioning) time-space:

1. Da-sein (being t/here as opening for disclosure).
2. Temporality (as temporalizing).

3. Be-ing.
4. Ab-ground.
5. Arising as such, holding-sway (*Wesen, Wesung*).
6. Enowning.
7. Thinging (gathering).
8. Saying (showing).
9. Opening.
10. Inconceivability.
11. Reservedness.

Though the tenth item on the list has not been referred to as a guideword up to now, it will serve as one from now on, to gather openness to mystery and releasement toward things together toward (the heart of) the attuning of the other beginning: reservedness. As Bokar Rinpoche said in another context in which mystery is in play for thinking and experience, "'Inconceivable' . . . means what it means."[12]

This reading of sections 238–42 of *Contributions* does not intend to cover everything significant in them but to move forward with enough to clarify the manifold transformations that are indicated and their grounding in/as ab-ground. Heidegger's thinking of time-space pulls together the thoughts carried in the guidewords listed above in such a way that we can begin to see (1) how we ourselves are pulled into the transformative sway of be-ing and (2) which questions must then be asked to move the matter forward. I will go carefully through sections 238–42 (not, however, in a strictly linear reading), thinking along with Heidegger, with added help from what I have already laid out, as well as closely linked texts (the ones that, as I pointed out in chapter 1, are in joining interplay with the thinking of *Contributions*).

To begin: "Space and time, each represented for itself and in the usual connection, themselves arose from time-space, which is more originary than they themselves and their calculatively represented connection. But time-space belongs to . . . the essential enswaying [*Wesung*] of be-ing as enowning. (At this juncture we need to understand why the point of reference of *Being and Time* shows the way in the crossing.)" (GA 65: 372/CP 259–60). We have already seen how the metaphysical, parametric concepts of space and time are derivative abstractions that flatten, respectively, thinging's gathering of sites or places and Dasein's temporalizing. Now, we need to be very careful not to misunderstand the significance of those moves in Heidegger's thinking. In the first place, this is not a matter of trying to obtain a more "correct" understanding, beginning with discrediting or falsifying the old concepts of space and time. "On the contrary, this [traditional] knowing will be above all relegated to the naturally limited sphere of its accuracy" (GA 65: 378/CP 264). This is not a call to bury our watches and clocks in the back of the sock drawer and forget about them (however appealing that notion might seem). Just as calculative thinking in general has its place *as long as its limitations are clearly acknowledged*, so too it is with the parametric notions of space and time that fostered and continue to accompany it. They are useful, but they are optional and, in fact, dangerous if they are allowed to determine the whole range of what is thinkable and how it may be thought.

The reference to *Being and Time* reinforces that clarification in another way and also pulls us firmly back into the context in which this thinking takes place: thinking in the crossing (and leap) into the first and other beginning. When *Being and Time* shows how "time" is an abstract flattening of Dasein's temporality, this is not meant to encourage us to (re)turn to some kind of "enriched experience" (*Erlebnis*) of temporality but, rather, to hint at something deeper yet: the time-space of be-ing (GA 65: 323/CP 227). And it is the thinking journey (experience as *Erfahrung*) into time-space that begins to hold the possibility of an other beginning more clearly open for us. The ongoing challenge is to try to bring this to language, which may seem to be "the usual representing" while it opens thinking beyond such representing conceptualization. Again, the point is not to abandon the old concepts and invent ones that might seem to us to be better but instead to dis-place our way of thinking and at the same time find ourselves being shifted into a more open arena (GA 65: 372/CP 260). We are not just juggling and rearranging concepts but instead undergoing a transformative experience with language. In section 239 of *Contributions* Heidegger says that our having thought the emergence of enframing (and the dominance of calculative thinking) from out of the first beginning's creating of "being" (and then "time" and "space") takes place as one way to (1) think this difficult matter using relatively traditional, easily understood language and (2) begin to be dis-placed into the open, having *begun* to overcome metaphysical limitations just in having thought them through even to that extent.

He goes on by pointing out that the thought of time-space emerges from this thinking of the first and other beginning. But, we might ask, what is the meaning of this word "time-space"? Heidegger cautions us that it is not to be understood as the coupling or linking of two separate entities or even of two distinct processes either in an ordinary, everyday sense or in the sense of mathematical space-time. Ordinarily, we think of time and space together in much the following way: all things are "somewhere" and "somewhen," and so they can be temporally and spatially determined. All this does is join together an everyday sense of parametric space and time. We might assume that contemporary relativistic space-time is closer to the time-space that Heidegger is trying to help us to think, but that concept takes time as another calculable element (the *t* of the equations), as a fourth parameter that, with the three spatial parameters, constitutes the four-dimensional space of physics. However, as Heidegger points out, this is essentially a very elaborate development from out of the original meta-physical flattening, abstracting, and grasping that create parametric space and time (GA 65: 378/CP 263).[13]

What, then, does Heidegger want us to hear and to begin to *think* in this word "time-space"? Another text gives us a clue: "Of time it may be said, time times. Of space it may be said: space spaces" (GA 12: 201/WL 106). Therefore, I suggest that we begin to call this guideword timing-spacing, emphasizing its dynamic implication: tim*ing*-spac*ing*. Then we can bring that together with what Heidegger says next in *Contributions*: "The one-fold of time and space is the onefold of origin" (GA 65: 378/CP 263). First, we remind ourselves that this is not origin thought metaphysically, as cre-ation or cause grounded on some being or explanatory first principle. That enables us, based on the joinings that I have already discussed, to bring together this much: this originary arising is the holding-sway (*Wesen*) of be-ing as ab-ground, whereby beings (things) come into their own (enowned in enowning). And now we can add: this enowning arising of things (the gathering of thinging) occurs as timing-spacing. Enowning, arising and holding-sway, and timing-spacing all *say the same*, belonging together as facets of the one matter: the be-ing of things, their dynamic, relational coming into their own. This already draws together much of what section 242 of *Contributions* is going to say and so will help us think farther along with Heidegger into it. In turn, doing so will clarify what I have just said.[14]

Our question is this: What does section 242 add to what has already been said? Obviously, that cannot simply ask for a neat division into what was said plus what can be added, as if it were a linear list of items. It asks us to heed what, in the enlarging joinings, (1) *says the same* in its differing nuances of meaning, so that (2) we will be open and able to be aware of the way in which the matter deepens and broadens here. It radiates out to "new" facets and then circles back, gathering them together in the onefold of be-ing and its saying (showing), which is to say, the same as "the one-fold of temporalizing and spatializing."

Timing and spacing, in their dynamic joining, are *inseparable* in their moving-together. As ab-ground they are utterly nonreifiable, and yet we cannot go to the other extreme and say that they are simply nothing (vacuous emptiness). "Something" is quite obviously happening, namely, a "manifoldness of enownings" (GA 65: 470/CP 331). This manifoldness says none other than the thinging of the thing, the showing forth (saying) of be-ing as enowning. Now, however, we are inquiring more deeply into *how* this happens. Thinging, we found, is gathering. So too, then, we can say that enowning is this same dynamic gathering, which happens (in terms of two other ways to say the same) as *turning* and as *timing-spacing*. Enowning as "relation-in-turning" says the way-making movement that "resonates back and forth" (instead of straight ahead, as if through metaphysical time). "Enowning has its innermost occurrence and its widest reach in the turning. The turning in enowning is the sheltered ground of the entire series of turnings, circles, and spheres" (GA 65: 407/CP 286). We know by now to hear this word "ground" differently; it is ab-ground, the staying away of ground that, says Heidegger, is also to be thought as *Ur-ground*. Ur-ground says primal ground*ing* that withdraws from any attempt to grasp at it through perception or conception. Not only does it withdraw in that way, in terms of language and thought, but this very withdrawing from *our* grasp intimates that its way-making movement *is itself a withdrawing*. Again, there is here a sense of resonating and circling in this enowning that itself withdraws, a coming forward that at the same time pulls away. But what comes forward into appearance is a thing, while what withdraws is its thing*ing*, its arising as such, the enowning that gathers and brings it into its own, to say (show) itself as this thing.

Enowning, enabling saying (showing), makes way in clearing and opening ways for gathering-thinging. Heidegger in another work calls on an Ur-word from the East, bringing it into play with the thought of enowning. That word is the Chinese word for "way," Tao, the "way that gives all ways" and that "makes way for everything" (GA 12: 187/CP 92). This clearly emphasizes the *dynamic* nature of enowning and, if we are acquainted with the work of Lao Tzu and Chuang Tzu, also serves as a powerful reminder of the nonreifiability of what we are trying to think. Listen to Chuang Tzu:

> To name Tao
> Is to name no-thing.
> Tao is not the name
> Of "an existent."

. . .
Tao is a name
That indicates
Without defining.[15]

This is why Lao Tzu, two hundred years before Chuang Tzu, begins his book of poetry on the Tao with these thoughts:

Tao is both Named and Nameless
As Nameless, it is the origin of all things
As Named, it is the mother of all things . . .
Tao and this world seem different
But in truth they are one and the same
The only difference is in what we call them[16]

Heidegger was not one to grasp at words just for effect. His bringing the Tao into the discussion of enowning and saying is very careful and deliberate. Furthermore, he knew enough of the original text (having worked with a native Chinese-speaking student toward a new translation in German) to have a good grasp of its meaning in its own context.[17] What Lao Tzu and Chuang Tzu say here echoes what has already emerged in our thinking of be-ing, enowning, way-making movement, thinging, and arising as such. If we attend carefully to what Lao Tzu says, however, we hear another thought as well: Tao and this world are one and the same. Just so, "be-ing is not something 'earlier'—subsisting for and in itself. Rather, enowning is the temporo-spatial simultaneity for be-ing and beings" (GA 65: 13/CP 10). Be-ing (enowning) and beings (things) are the same and simultaneous, which does not say that they are identical. "Same," we know already, means here that they belong inseparably together. What about this temporo-spatial simultaneity? What does that say? First of all, it tells us in yet another way that be-ing is not a being, much less one that could somehow exist *before* the arising of beings. But to say more we need to leap into the discussion of timing-spacing's way of moving, wherein "time and space, in all strangeness, are grounded in their originary *belonging-together*, [for] clearing and sheltering . . . proffer[ing] the transposed open for the play of a being" (GA 65: 69/CP 48).

I will approach that by putting forward one of the passages that attempts to say how time and space belong and move together, but, as Heidegger indicates, it will at first sound very strange or very nearly impenetrable.

Space is rendering ab-ground that charms-moves-unto the gathering.

Time is rendering ab-ground that removes-unto the gathering.

Charming-moving-unto [*Berückung*] is the encircling hold of gathering that holds to ab-ground.

Removal-unto [*Entrückung*] is gathering unto the encircling hold that holds to ab-ground.

Here we have some familiar touchstones, ab-ground and gathering, with the introduction of the strange-sounding "charming-moving-unto" and "removal-unto." What does this say? Drawing on the context of this page in *Contributions* (GA 65: 385/CP 269), with help from Heidegger's attempts to say the same in "Time and Being" (TB 16–20) and *On the Way to Language* (GA 12: 199–203/WL 105–7), we can carefully work toward understanding.

This "moving-unto" and "removal-unto" offer more specificity as to what goes on in the turnings in enowning, whereby things are gathered and brought into their own. What is added here is the *inseparable dynamic* of temporalizing and spatializing. Notice that both are said to be gathering, that is, this is the *one* movement of the thinging of the thing, but it is a complex motion (resonating, circling, intertwining) that opens and extends, clearing and making way for *and as* a thing's arising. The moving-unto says something of spatializing, and the removal-unto says something of the temporalizing of the thing. That *removal* is emphasized tells us that in this gathering there are also "*dis-placings* of merely *what*" comes forth and is brought to light. What does this mean? Back in chapter 1 I said, "Change one thing, change everything." That is true not only of thinking but also of the intrinsically relational gathering that is the thinging of the thing. Thinging, in its own way, is a matter of joinings. The bottom line here is *continuous change*. It is not just that be-ing is not a being. At this point I also have to say that *a being* is not a being (something that is simply present as *what* it is). "A being" is now thought as a thing that arises dynamically, relationally, and that never quite stands still in presence. There is presenc*ing* (be-ing) but not presence (being).

The nuance that "charming" (or attracting, or enchanting) adds to the moving-unto that gathers and spatializes is an indication that this is not just some random, chaotic free-for-all. Certain things and processes tend to pull together more than others do. At the same time, this is never simply sweetness, light, and harmonious joining. In the parallel account in "The Thing" we read that the "nestling, malleable, pliant," and compliant play of thinging also "*wrests* [*entringt*] free the united four[fold]. . . . Out of the ringing mirror-play the thinging of the thing takes place" (GA 7: 182/ PLT 180). *Ringen* and *entringen, ereignen* (enowning) and *enteignen* (disenowning), charming-moving-unto and removal-unto, assonance and dissonance: the thinging of the thing. No wonder it never stops stock-still, never solidifies into what is simply present, simply a being (see also GA 65: 260–61, 472/CP 183, 332).

What do we have thus far? Moving-unto says something about spatializing, and removal-unto says something about temporalizing. And both say something about the way that the gathering into a thing takes place, the way that the turnings in enowning are enacted as *counterturning*. But we must be careful not to let this resolve into a duality, like a simple moving together and moving away. That caution arises from two sentences that might be the most difficult to understand of anything that Heidegger ever wrote: "Time spatializes [*räumt ein*], never charms-moves-unto. Space temporalizes [*zeitigt ein*], never removes-unto." What does this say? Most likely, it says more than what I can encompass here, but we can nevertheless gain some clarity. Removing-unto is a pulling away that at the same time gathers; this says the *changing* that is the movement away from what something "was" to what it "will be," a gathering to that is at the same time a dispersing from. Gathering thus cannot and does not solidify into "a being." This "dispersing," however, is *always* in play with moving-unto. Thus, the dispersing (temporalizing) also spatializes in opening and clearing the way for moving-unto. And it—the dispersing—cannot and does not simply scatter out into "empty space." So, as Heidegger said, "time spatializes." But that sentence ends with "never charms-moves-unto." The two, spatializing and temporalizing, are inseparable and essential to each other but do not simply collapse into one another. We speak of "one" and "two," but this is utterly outside the realm of any question of philosophical monism or dualism. Timing-spacing is not *one,* nor is it *two.* "Inconceivable" means what it means. Anything we can say about this will only be partial, will only capture a few facets of Indra's net.

I thought that far in terms of "time spatializes," but, as we might well assume at this point, Heidegger also said, "Space temporalizes." Can I articulate that more clearly? The pulling toward and together (charming-moving-unto) that gathers the thing, making a site or place for it, is, once again, a way to say *changing*. It is a changing that opens and extends, that makes way, in its wake and in its neighborhood, for a shifting in the relational web. What was is no more (never to return and yet somehow still in play), and what will be is only just and/or not-yet happening. So it is that spacing temporalizes, just as timing spatializes. Timing and spacing name distinct aspects of the turnings in enowning yet always remain inseparable as they enact the gathering of thinging (see also GA 65: 191–92/CP 134–35).

Another word of caution is appropriate. The structure of our language requires us, even in attempting to think something like this, to use grammatical constructions in the subject-predicate form: "timing does X," "spacing enacts Y," "enowning turns," and so on. So even though we keep in mind the nonreifiability of any of these words of be-ing, it remains difficult to hold back from thinking of these guidewords as *something* in and of themselves (TB 16). In section 242 of *Contributions* Heidegger brings in two more ways of vividly reminding us of the nonreifiability of be-ing (by whatever name). The first is the thought of the *uniqueness* of be-ing. "This essential sway of be-ing is unique and once only . . . φύσις (*physis*) too is unique and once only." The first beginning's wonder at the always-unique arising (*physis*) into a being eventuated into the creation of what in hindsight we can call the ontological difference, which flattened the each-time-unique arising into what is common to all beings, their presence. To think *that* in its power as beginning the whole history of metaphysics opens the way to now leap beyond its limiting hold and think be-ing as the always-unique gathering-arising of a thing. *Apart from thinging there is no be-ing.* That is why be-ing is said to be "always-unique and once only."

This thought of the uniqueness of be-ing also disrupts the question of whether be-ing is infinite (as being was said to be) or finite (as a being). While "finite" is perhaps closer to the mark, be-ing is neither finite nor infinite because both of those intrinsically metaphysical concepts pertain to being and the beings it grounds. Furthermore, finitude and infinity are concepts that developed along with parametric time. They have to do with the *count* of nows that assign quantity to time. That says nothing relevant about timing-spacing. To say that the uniqueness of be-ing is finite would be to flatten its dynamic relational gathering. To say that be-ing is infinite also does not go beyond the flattening; it just imagines an endless series of flattened now-points. The *unique moment* of timing-spacing is so much richer than the now-point of "time" that there is no way to reduce the former to the latter except by arbitrarily stopping its dynamic in order to reify it into a set of concepts. "Only when something extant is held onto and fixed does the flow of 'time' that flows by the extant arise, only then does the 'space' that encompasses the extant arise."

The other way that Heidegger reinforces the thought of the nonreifiability of be-ing in section 242 is by introducing another way of naming be-ing as ab-ground: emptiness (*Leere*). This word enters into the discussion several times, but I have for the most part carefully worked around it because I am holding back a fuller discussion of emptiness for chapter 5. Here, however, I can say in a preliminary way that emptiness is not vacuous nothingness, nor does it have anything to do with any failure of our expectations. It is in play, rather, as another way to say the nonreifiable fullness of the dynamic, relational, timing-spacing that *opens and clears* the way for the thinging of the thing. It is the very possibility of thinging.

After all that, how do Heidegger's comments at the beginning of this chapter sound? I am thinking in particular about this: although the relations of timing-spacing are inaccessible to calculative thought, they are actually simple if we can properly attend to things in their mysterious richness. Remember, "simple" here does not mean "easy." The language of the texts in which Heidegger attempts to think through timing-spacing remains, as texts, very difficult. But what he said about the simplicity of these relations does not, in fact, refer to texts but to timing-spacing-thinging.

Return once again to that loaf of bread. In gathering earth, sky, mortals, and the intimation of mystery in it all, this loaf comes to stand in this place on the counter, in this momentary uniqueness of be-ing. In this gathering what Heidegger calls charming-moving-unto tends to pull together wheat and yeast and oil and water rather than centipedes and shoestrings and potatoes. In this movement, the spatializing of this thing, the wheat and yeast and all the rest of the things in play are changing. Wheat becomes flour becomes bread becomes sandwich becomes carbohydrate in my body. It is not all that difficult to see here at least some of what Heidegger means by removal-unto, which speaks of the temporalizing of the thing. And, of course, if we stay with this loaf and do not wander off into abstractions, it is also clear that the temporalizing and spatializing, while naming distinctly thinkable aspects of the gathering, are inseparable. Without moving-unto there would be no loaf. Without removal-unto there would be no loaf. Pulling that together: without timing-spacing there would be no loaf. But also—and this is crucial—without the loaf, without *thinging*, there would be no timing-spacing. So it is not only timing-spacing that is inseparable but timing-spacing-thinging.

There is one more significant trail of thought in section 242 that I have yet to touch on. That is this question: Where are *we* in this timing-spacing-thinging? "Ab-ground is thus the in-itself temporalizing-spatializing-counter-resonating site for the moment of the 'between' as which Dasein is grounded." This comment is very nearly at the end of section 242. It reminds us that we, too, are in the midst of the relational dynamic of timing-spacing-thinging. We arise within it and in that way are akin to all things. Furthermore, we are be-thinged (*be-dingt*) or conditioned by them (GA 7: 182/PLT 181). But we do not, nevertheless, tend to refer to ourselves as things. What is ours in particular is (1) to be Dasein, to be the t/here as opening for disclosure (*Being and Time*) and (2) to respond to the pull, the call, of the saying of things so as (3) to think them and thus (4) to dwell among them as those who tend and preserve things rather than those who, in flight from thinking, ignore and destroy them. That is the challenge to which chapter 4 attempts a response.

4

Thinking as Dwelling

Heidegger concludes "Building Dwelling Thinking," the essay in which he discusses the thinging of the thing and the derivative character of parametric space, with these words: "Think for the sake of dwelling" (GA 7: 164/PLT 161). To understand this and take it seriously requires broadening our notion of what it means to think even more than we already have. So far, thinking along with Heidegger, we have learned at least the following things about what he means by this word "thinking." Our typical ideas about what good thinking amounts to have been shaped by the long history of Western philosophy in play with popular culture. In opening up the thinking of the first and other beginning, including the transitional thinking in *Being and Time*, we learn that our most deeply (so deep as to be hidden) held assumptions about "being" are the product of the creative thinking of the ancient Greeks in response to their wonder at beings' arising and coming to presence. Bringing that wonder into play with *technē*, understood as our ways of preserving the being of beings in their relation to us, they first conceived of what we, in hindsight, may call the ontological difference. Thus the being of beings begins to move apart from beings, at first conceptually and then (as the history unfolds) as an item in Western ontology (as, in fact, the very basis of metaphysics). Unthought by the Greeks themselves in terms of its being a *creative* move in thinking, this very splitting apart of being and beings is forgotten, setting in motion a complexly unfolding web of consequences by which we are still shaped and constrained, especially now at the extremity of metaphysical thinking as enframing, whereby beings begin to lose any unique standing at all (even as mere objects) and become more and more just interchangeable units in a reserve or stock of resources on call for production. We ourselves, who presumably are in charge of this productive framework, stand on the brink of also becoming nothing more than stock and of being able to think in any other way than calculatively. However, in even being able to think this history and our place in it (to think, as Heidegger calls it, the first and other beginning as such) its power to limit us is loosened; we are already under way within the opening toward an other beginning.

To proceed, according to Heidegger, requires at this time that we mindfully learn what thinking has been in order to unlearn it. This does not mean simply rejecting calculative thinking but, rather, learning to let it be optional in the same way that Heidegger suggests we comport ourselves toward the products of technology, being able to say "yes" or "no" to them according to the actual context. This he calls releasement toward things. In terms of enacting this releasement toward the "things" of metaphysical, calculative thinking we learn to let go of the constraining assumptions and elements of traditional Western thinking: the necessity of method, rules of logic, definitions, and even concepts. We learn that these will not serve as we attempt to reawaken our wonder at the arising of beings and think this arising as be-ing, that is, as ab-ground, which in no sense is or has "being" (reifiable presence). How can we *think* something like this or even know how to hold it in question? Taking up Heidegger's hints in "Memorial Address," I suggested that we approach the matter by way of both releasement toward things and openness to mystery. As I have discussed these so far, they mostly have had to do with our relationship to and use of language. Releasing has been applied, as just mentioned, to the ways of thinking determined exclusively by the first beginning and its most extreme development, enframing. That is essential right from the start in order to even begin to be able to engage with the mysterious holding-sway of be-ing. *Mystery* has been discussed so far mostly in terms of (1) the nonreifiability and inconceivability of be-ing and (2) the way that timing-spacing-thinging can be both so simple and yet so astonishingly difficult to bring to language. In anything we say it seems that something is just out of reach, slipping around the corner just in front of us.

So it is that we learn to work with guidewords, which echo timing-spacing-thinging in its dynamic relationality. How so? Guidewords are open in that they are not definable and are not even "concepts" in that they do not decisively grasp and fix what they say. They have multiple meanings that shift as the context shifts, but they all nevertheless "say the same" (be-ing) in different ways. Guidewords enact their saying (which is, as showing, the heart's core of language) without reifying, reducing, limiting, or halting thinking. At this point I note again the crucial importance of Heidegger's observation that saying (showing) does not belong only to language but to all things. That, indeed, is why language can say anything at all in responding to the saying of things. The challenge for this attempt to think things in their thinging (as their enowning in timing-spacing) is to remain attentive to letting the language continue to echo the originary saying of thinging without yielding to the temptation to rest on some (always partial and tentative) saying as "the last word." This is why the word that shows up most often in *Contributions* to designate the manifold attuning of transformative thinking is *reservedness*. Reservedness (1) resonates with and is pulled along by the withdrawing of be-ing and (2) stays open to saying's way-making, (3) following the movement of thinking rather than forcefully pouncing on words or ideas along the way. To stay with that demeanor in thinking is to bring together openness to mystery and releasement toward things in our relationship to language. That is already a significant step, this being willing and able to undergo "an experience with language . . . [that] overwhelms and transforms us" (GA 12: 149/WL 57). However, the transformative thinking of the first and other beginning calls for more yet. This "more" has already been hinted at in the insight that the saying of language resonates with the saying (showing) of things. In the preface to this book I raised the question, What is this thinking *for?* That question does not have just one answer, but it would seem that when Heidegger says, "Think for the sake of dwelling," we may well be on the trail of a response.

There are two misunderstandings that should be cleared from the path right away. The first is the idea that "dwelling" is the desired end or result of thinking. We are fairly well prepared to see why that would be a misunderstanding. The possibility of dwelling opens up within the transformative thinking of the first and other beginning as it goes deeper into the thinking of timing-spacing-thinging. We already know that an *other* beginning is not thought as the end result of thinking the first beginning. Instead, the possibility of an other beginning already opens within the thinking of the first beginning as such in its power *as* beginning. Another way to say this is that what we might be inclined to call "preparatory" thinking is *already* transformative. This understanding is reinforced by the insight that this thinking and the language that carries it—especially as sparked and moved by the joinings of guidewords—is simply not reducible to a linear, premise-and-conclusion, step-by-step process. So it should come as no surprise that a *sequence* like (1) preparatory thinking, then (2) transformative thinking, and then (3) dwelling is not how this matter unfolds. The relation of (1) and (2) has already been discussed. What is in front of us now is to attempt to understand the relation of the third item, dwelling, to the first two. I will say this much now: Heidegger's saying "Think for the sake of dwelling" does not imply a cause-and-effect relation between thinking and dwelling. Neither does it suggest a subordinate, means-and-end relation, that is, that thinking is merely a means by which to accomplish dwelling. Thinking and dwelling belong together. Heidegger tells us that we must think if we are to dwell (GA 7: 163/PLT 160–61). I will strongly urge—by the end of this chapter—that this necessity is reciprocal: without dwelling, thinking is impoverished, and we are returned to the uprooted thoughtlessness that goes with exclusively calculative thinking as described by Heidegger in "Memorial Address."

Coming to grips with the nature of that misunderstanding of the relation of thinking and dwelling prepares us to deal with another misunderstanding. "Dwelling" is not the name of some utopian vision or of Heidegger's notion of an ideal way of life for humanity. When thought—as it must be—in the context of the first and other beginning, dwelling is thought as the possibility of a shifting of human be-ing arising from and with transformative thinking. The possibility of dwelling first emerges from within our undergoing a transformative experience with language; that already says a transformation of thinking *and* of our way of interacting with things. Language—as the "relation of all relations" for us humans—both reflects and shapes our understanding of and relationship with things. It is clear that dwelling stands in quite distinct contrast to the exclusively calculative thinking that enacts enframing. For many of us, as we begin to understand what this means, dwelling also sounds *better*. So be it; I would be the last to contradict that. However, here again it would be a mistake to take any kind of theoretical, either-or, black-and-white stance. For one thing, we do not yet know sufficiently what it means to dwell, or how to go about it, such that we could or should simply replace "calculation and enframing" with "thinking and dwelling." Calculation is a limited sort of thinking, after all, and in its own place it is not necessarily harmful. Furthermore, this is, in any case, not a matter of simply replacing one set of such concepts and norms and actions with another. This dynamic web of transformations is much more complex than that as well as more tentative and indefinite. What dwelling means will only emerge within that larger context, in play with the transformation of language and thinking, in response to abandonment of being, open to ab-ground, and attuned by the reservedness that both springs from and enacts the thinking of the first and other beginning. As the thought of dwelling begins to emerge more clearly, it in turn will shift the net of joinings, impacting our understanding of language and thinking.

STAYING WITH THINGS

In chapter 3 we saw that, unlike a being, a thing is to be thought dynamically, as *thinging*, which means a gathering of the whole web of relationships into what emerges as its own (enowning). We then thought deeper into that matter in terms of how timing-spacing enacts openings that clear the way for thinging. Even that sentence, accurate though it be, is a bit too linear. Heidegger is quite clear that timing-spacing-thinging is *all at once* and unique. Heidegger brought the web of relationships to language poetically as the fourfold of earth and sky, mortals and divinities. We, the mortals, are always already in the midst of this timing-spacing-thinging, just as things are. Toward the end of the discussion of space in "Building Dwelling Thinking" Heidegger provokes our thinking to go farther along that path in saying that I am not just "here" in one parametrically determined location "in this encapsulated body; rather I am there, that is, I already pervade the room, and only thus can I go through it" (GA 7: 159/PLT 157). This accords with our understanding of both temporality (as explained in *Being and Time*) and timing-spacing-thinging. "Here" and "now" are emerging as mere shorthand names for aspects of dynamically relational timing-spacing thinging. That is not at all difficult to see in terms of things. They are not beings, that is, they are not reifiable entities that can be thought in terms of their being grounded on some interpretation of being (constant presence). But Heidegger also says that *we* too are not just "here" and "now." That hints at a thought that many of us may find rather alarming: *we are not reifiable entities* either. I put that thought forward now so that we can hold it in front of us as a question (What does that mean? And what are we, then?) during my discussion of dwelling.

The German word for dwelling, *Wohnen*, has its roots in older words that mean to remain or stay in a place, to be at peace there, and, as we might say, to be at home there. The peace of being at home suggests a caring for the things among which we dwell, not harming them, and, more strongly, preserving and even freeing them. Heidegger pushes this farther, suggesting that only if we can free things will we also be free (GA 7: 150–51/PLT 148–49). Here we encounter an additional nuance of meaning for one of our guiding thoughts: releasement toward things, which is now put in play with the thought of "the free" and freeing. What does this mean? As we might expect, that is difficult to simply say directly but must emerge gradually. It does, however, give a strong hint as to the direction of the thinking.

What follows immediately after that comment on "the free" in "Building Dwelling Thinking" is the account of thinging that I have already laid out in chapter 3. Then Heidegger begins to suggest how we might begin to think the nature of dwelling, which is to spare, preserve, and free the fourfold. That means (1) saving the earth, refraining from attempting to master and subjugate it, (2) receiving the sky, or being aware of and in tune with the primal timing of day and night, the seasons, and so on, (3) awaiting the divinities by not reifying them (inventing gods, idolatry), and (4) learning to live as mortals, neither evading our mortality nor becoming nihilistically obsessed with it. While all of these are relatively easy to comprehend, they seem a bit *large* if we were to take them on as tasks of some kind or even as demeanors. The fact is that we never actually interact with "earth," "sky," "divinities," and "mortals" *as such*. They are always encountered in their inextricably relational dynamic: timing-spacing-thinging. That is why Heidegger says, "Mortals would never be capable of it if dwelling were merely a staying on earth under the sky, before the divinities, among mortals. Rather, dwelling itself is always a *staying with things*" (GA 7: 153/PLT 151, my emphasis). So we dwell with what we actually encounter, namely, things, by *staying* with them. How? The first response from Heidegger is that it is to "nurse and nurture the things that grow, and specially construct the things that do not grow" (GA 7: 153/PLT 151). That sounds quite straightforward and down-to-earth. It is, and yet this says much more than a quick first glance will notice. A hasty response would be that we are already doing this. We garden, we farm, we raise children and other animals; we are continually constructing and building and crafting and manufacturing and producing a seemingly endless stock of things. But this hasty response, based on that simple statement of Heidegger's, overlooks one crucially important detail: dwelling is staying with *things*, nurturing and constructing *things*. As denizens of the world of techno-calculative enframing, when we hear or read that little word "thing" we still tend to hear and understand it as meaning more or less the same as "a being." And beings we understand, through education and application, as things having such and such properties, useful for such and such purposes, and thus either fitting into our worldview and "lifestyle" or not. But, as Heidegger reminds us, "our thinking has long been accustomed to *understate* the nature of the thing" (GA 7: 155/PLT 153). Having just attempted to think the thing as timing-spacing-thinging, we know from where that comment comes.

The fact is that *staying with things* through nurture and construction is nothing at all like what we are accustomed to in our contemporary techno-calculative world. The German for "to dwell," *wohnen*, is closely linked to a group of related words (*gewöhnen*—a verb—and its noun and adjective forms) that all have to do with habit, custom, what is usual, and what we are used to. We have this kind of connection in meaning in English, too, in that our word "inhabit" has as its root "habit." To become habituated to a place or its customs is to learn to dwell there, to become acclimated to the habitual ways and customs of the place, to get used to them. So what sounds so down-to-earth that it seems obvious and easy—staying with things through nurture and construction—is actually pointing toward deeply rooted and far-reaching transformation. We are accustomed to relating to beings within a metaphysical, calculative framework. We are thus habituated to a certain set of understandings and ways of dealing with be-ings. Dwelling calls for shifting, for reinhabiting a world of things emerg-ing in the ab-ground of timing-spacing-thinging. To even attempt to thus stay with things requires first of all *remembering to think them as things*. All three facets of that phrase are crucial: remembering, thinking, and things.

THINKING: THANC

In the domain of enframing not only are things of little account. *Mystery* is either absent or taken as an indication of a problem in need of a solu-tion. And to associate "thing" with "mystery" would seem very odd indeed. But, as I have suggested, things are, in contrast to the way we usually just take them for granted, deeply mysterious. They are the most ordinary and, at the same time, utterly extraordinary. This, of course, is because they are, in and of themselves, the open, relational dynamic of timing-spacing-thinging. There is thus in simple *things* an inexhaustible depth and breadth and energy. So when Heidegger says that dwelling means to stay with things so as to preserve what is their own we face quite a challenge. Because the thing is inexhaustible this is not a problem to be solved, so calculative think-ing is of no use whatsoever. And if we were to "figure something out" about things and rest satisfied with that (thinking that now we have grasped the truth about things), we would merely have returned to a "being" rather than staying with the thing. It matters not which word we use—it is not, after all, as if there is something wrong with the word "being"—but, rather, how we are thinking of the matter. As Heidegger puts it, "We attain the simple only by preserving each thing, each being, in the free-play of its mystery and do not believe that we can seize be-ing by analyzing our already firm knowledge of a thing's properties" (GA 65: 278–79/CP 196; see also GA 12: 241–42/WL 122; GA 5: 41/PLT 54; GA 65: 10, 108–9, 131/CP 8, 75–76, 91). We are, in a way, in a similar situation to that of the ancient Greeks, who first realized that they simply *did not know* how to think and say the arising of beings or just what the being of a being "is." This struck them as astonishing and moved them to wonder at what is uncanny in the most ordinary of beings. It was this wonder that attuned their questioning (what is this, the being of a being?) and thinking. Now, in the midst of the emerging thinking of the first and other beginning, we, too, do not fully know how to think and say *things* or even how to dwell with them, staying with them, preserving and freeing them. Wonder, once again, seems an appropriate response. But where does it take us?

It is becoming apparent that even all of what we have learned so far about noncalculative thinking is not yet adequate if we are to stay with things, to think them and dwell with them. We need all those insights about the *saying* of things, emerging also as what is ownmost to language. But there is, it seems, something missing.

I recall that in "Memorial Address" Heidegger said that any of us can think if we begin by dwelling on and pondering what lies close and deeply concerns us (DT 47). Certainly, among many possibilities, the things all around us are close and even more than close: intimate. We are not just an encapsulated body, and neither is a thing just what we see in its bounded shape. We are *in* the thinging of the thing; things are, in that same way, *in* us, the be-thinged (*be-dingt*, conditioned) ones. Whether we give that a thought or not, it is so. Assuming we do want to think and dwell with things, how do we "give it a thought"? Let me alter that question slightly: What is it (in us or as us) that thinks?

We are, in consequence of our long-standing philosophical tradition, used to a clear and unhesitating reply to that sort of question: why, it is— of course!—*the mind* that thinks. But surely by now we know to hesitate in simply accepting without question the dualisms that have shaped Western thinking, particularly the subject-object and mind-body dualisms. We can certainly continue to use the word "mind," but without body—participating in timing-spacing-thinging—what would mind "be"? And on what basis would we—other than merely as a conceptual abstraction—*separate* mind from body? I have no good reason whatsoever to reify myself—or any part or aspect of myself—any more than to reify things. So if "I" am not a being, how could my "mind" or "body" somehow name beings that could be separated out as distinct entities? We can question similarly regarding "subject" and "object" (a matter that is taken up again in chapter 6).

So I ask the question again: What "is" it that thinks? But perhaps that asks too much or asks in the wrong way. Try again: What "is" it, in me or about me, that impels me to think and that carries out thinking? Here I turn for guidance to what I take to be one of the most crucially important, powerful things Heidegger ever wrote. This is the passage on the *thanc* in *What Is Called Thinking?* Many readers of Heidegger are aware of this section of text. Occasionally, someone refers to it in print, tossing out a platitude such as "thinking is thanking." There is, however, something much more profound going on here, something with powerful ramifications for the entire quest of thinking in the first and other beginning. From here forward I use this word thanc—just so, with no italics or quotation marks— as a guideword in joining with all the other guidewords.[1] What follows is a carefully considered interpretation of pages 139–51 of *What Is Called Thinking?* (WHD 91–97, 157–58), bringing them into play with the thinking of be-ing toward an understanding of what it means to say that dwelling is *staying with things.*

As he often does, Heidegger opens us to a different way of thinking about the meaning of a key word by drawing our attention to nuances of meaning carried in its past usage in language. Here, he begins with the word "to think" (and its German cognate, *denken*). Focusing more on the English, he notes that when we go back to the Old English, the words that then meant "think" and "thank" begin to converge, especially in the noun *thanc*, which meant a thought and, particularly, a thought or expression of gratitude. He takes this as a clue to open up the region from which and within which thinking moves. As is the case with all the guidewords for thinking, thanc is not simply definable, so it is not limited to the "grateful thought." It also has a sense of recalling and remembering, as a *gathering* of what is thought and what is to be thought.

Thinging: gathering into things: timing-spacing.
Thanc: gathering of thinking, with recalling (thus timing is intimated).
Saying: the showing of both things and language.

This joining suggests that we are moving in a direction that will open a way into the matter in question. And there is more yet to thanc.

The word, as Heidegger works with the older meanings and thoughtfully brings them into play with the question of the book (What is called thinking? What calls for thinking?), says something not only about what we do (think) but about what inclines us to think and, indeed, motivates us so deeply that it shapes our very humanity. "Thanc" already suggests a focused concentration on what concerns us most intimately. Going deeper, Heidegger says, "The *thanc* means [our] inmost mind, the heart, the heart's core . . . [that] reaches outward most fully and to the outermost limits, and so decisively that, rightly considered, the idea of an outer and an inner world does not arise. . . . The *thanc*, the heart's core, is the gathering of all that concerns us, all that we care for, all that touches us insofar as we are, as human beings. . . . In a certain manner, though not exclusively, we ourselves are that gathering" (WHD 157/WCT 144). There are at least two things said here that require our careful attention.

First, the thanc, while we may call it mind (since it impels thinking), is not "mind" in the limited way in which we usually understand that word. At this point this is not surprising, coming after much talk of what deeply concerns, touches, and moves us. Notice also that use of the phrase "the heart's core." So thanc is neither just mind (intellect, reason) nor only heart (affect, feeling). It is closer to something that we could, at least tentatively, call heart-mind. I think it is also important to note here that when we think of what is—ordinarily and also here, with Heidegger—meant by "heart," the body is also implicated. The farther we go with attempting to think be-ing, to think things (timing-spacing-thinging), and also to "think thinking," that is, to understand just what it is that we are attempting, the more difficult it becomes to hang on to old assumptions about ourselves, about what it is to be human in any deep or rich sense. I can now venture a preliminary response to my question (What is it that thinks?) and say that it is the thanc, our heart-mind-body. It is the thanc that gathers and keeps what is to be thought. Recalling, remembering, is held within the body, just as we thought along with Abram in chapter 3, in seeing one way that timing is enacted and even perceived in our embodied relationship with things. It is held, kept, dynamically, arising sometimes without our intending it and other times to be called upon with intention. In turn, what is to be thought also extends "to the outermost limits." Well, of course it does: the thing gathers all, in its timing-spacing-thinging.

At this point we have to look more clearly at Heidegger's remark, "We ourselves are that gathering." What gathering? The gathering of what is to be thought. But that, in turn, is the gathering that is timing-spacing-thinging. And, in some sense, we are that gathering, too, not only the gathering that we identify with the word "thinking." I have suggested already, earlier in the book, that our notion of ourselves as *beings* is going to become more and more questionable. Here is another indication that we are more akin to things (in terms of the *relational dynamic* of thinging) than we are to beings (something present, something substantial, that is, self-existing). I am not going to leap to any conclusions about this at this time, but we need to be aware of the hints that are gathering all around that issue and that will come to a head in chapter 5. Here we note that our heart-mind-body gathers thought in response to the gathering of things, the things that are of deep and intimate concern to us. They are so intimate, in fact, that in some sense they "are" us. As Heidegger puts it, at some point "the idea of an outer and an inner world does not arise." Part of what this gathering-thinking involves is named "memory," the gathering of what was and is unfolding (by now we know that this is not a reference to linear time) (WHD 95–96/WCT 149; see also GA 65: 257/CP 181–82).

What about the Old English meaning of the word "thanc" that Heidegger used to open the discussion? Where does gratitude or thankfulness enter meaningfully into thinking? I would suggest that we let it join with openness to the inexhaustible mystery of timing-spacing-thinging and the wonder that arises from that openness. If that begins to seem too huge, somehow, to sustain, remember what Heidegger said about dwelling. We would not be able to do it if it really had to be carried out as saving the earth, awaiting the sky, and so forth. We do not interact with the fourfold as such but with things. Likewise, even though we and things alike "are" timing-spacing-thinging, which is thus the grounding of our intimacy, it is hard (and perhaps even impossible) to imagine *dwelling* (staying) with that as such. So it was that Heidegger said that dwelling is staying with *things*, to think and tend and nurture them. In terms of the thanc, too, we are mindful of the largest possible relational dynamic (the "outermost limits"), but what actually moves us, evoking wonder and, yes, gratitude, is that this immense, mysterious enowning of things and of us manifests in things, in what is so close and ordinary. They are so close that, in addition to the word "intimate," Heidegger says that they are "contiguous" with us, they touch us; in touching us, they—in the other sense of touch—*move* us to stay with them attentively. This devoted, attentive staying is thinking, it is dwelling. It is not all there is to thinking and dwelling. But it is their heart's core and the necessary opening toward being able to carry them out.

How, then, are we to carry forward with thinking and dwelling? Heidegger, in one turn of the discussion of this theme, suggests that we might give thought to our manner of *hearing*, which is decisive for understanding thinking by way of the thanc (WHD 158/WCT 145). In chapter 2 I pointed out, in discussing the understanding of language that is needed to attempt to think be-ing, that Heidegger plays on the link between the German words *hören* (to hear) and *gehören* (to belong). There, the core thought was that because we belong within language we can properly hear and heed what is said (saying-showing). But now the scope and complexity of our belonging have been opened up considerably, to the extent that our intimacy with things, in the play of timing-spacing-thinging, pervades us through and through. That should give more depth to what Heidegger said about saying: it not only is the heart's core of language but also names the showing or display of all things. At this point it can begin to seem that there are so many facets, so many threads coming together as "inseparable," "belonging together," and "saying the same," that it becomes difficult to hold the thought. On the one hand, that is to some extent both unavoidable and meaningful, insofar as it is another way of being reminded of the withdrawing of be-ing and saying: mystery comes to meet us yet again. On the other hand, it is important to not let this wide-ranging, multifaceted gathering become so slippery that it ends up as just another academic exercise. So there are two things to take up that will help us avoid that: (1) bring more of what was said earlier of language and thinking into play with the thanc and then (2) return to the question of staying with things in their "ordinary," everyday uniqueness.

Language and the Thanc

In chapter 2 there was one thing that was left a little vague because there was at that time too little that had been said to serve as a basis for explaining it more clearly. That was the notion of thinking as *in-grasping*. Heidegger contrasts this with concept formation (the move from *begreifen*, understanding that grasps, to *Begriff*, the concept, the mental "grasp"). Several reasons were given as to why the thinking of the first and other beginning is not well served by such grasping. The concepts that we acquire or construct limit or fix thought and tend to effectively hinder or even close off questioning. But insofar as words do "have meaning," it seems that there must be some kind of grasping involved in any use of language. But perhaps what is grasped can at the same time be released and freed or, in a

sense, grasped with a light touch. This differs from theoretical, propositional grasping and instead is the "'in-grasping' [*Inbegriff*], and this is first
and always related to the accompanying co-grasping of the turning in
enowning. . . . In-grasping here is never a comprehensive grasping in the
sense of a species-oriented inclusiveness but rather the knowing awareness that comes out of in-abiding and brings the intimacy of the turning
into the sheltering that lights up" (GA 65: 64–65/CP 45–46). What I was able
to say about that in chapter 2 was that the turning in enowning says the
dynamic of be-ing and that in-grasping, unlike conceptual grasping, is
attuned by *reservedness*, in awareness that be-ing, as ab-ground, always
withdraws from our attempt to grasp it in thought and language. I also
suggested that as we, too, are enowned into be-ing we do not just grasp
something "out there" but also "in here," hence, *in*-grasping. That also hints
forward to the thought that the barrier between subject and object, and
even between inner and outer, would fall to the wayside, decisively transforming our understanding of ourselves. With, for the most part, only a
discussion of thinking and language as context, that may well have sounded
like a mere rearranging of words ("grasp" to "in-grasp") without much else
to go on other than the joining with the guidewords in play at that point
(be-ing, enowning, etc.). Now the discussion of timing-spacing-thinging
in chapter 3, along with what has just been said of the thanc, should open
up a clearer sense of what in-grasping means. We can in-grasp what is said
(shown) because it already resonates *in* us, too, in the dynamic of thinging, in which we are always already involved. This in-grasping of what is
said does not have to be articulated in spoken or written language, any
more than the saying of things always manifests in sound. The last paragraph of *Contributions* says that "language is grounded in silence . . . as
essential holding-sway of the jointure and its joining" (GA 65: 510/CP 359,
translation altered slightly). The silence of things, of earth and sky and
divinities, is not dumb. It *says* or shows many, many things. In its saying,
which echoes and resonates in us as what Heidegger calls the "ringing of
stillness," it calls on us to listen and to respond, to hear and to heed from
our heart's core, the thanc, from our belonging within timing-spacing-
thinging and its saying (GA 12: 204, 244–49/WL 108, 124–30; GA 12: 26–29/
PLT 206–9; GA 65: 56–57/CP 39–40). Heidegger uses two different words to
say our way of thoughtfully responding to the saying of things: *entsprechen*
and *nachsagen*. *Entsprechen* generally means "to correspond or to be in
accord with something." Its root, *sprechen*, is "to speak." A glance at any German dictionary tells us, by way of pages upon pages of words, that *ent-* is

a very common prefix, with several meanings. Here it carries a sense of *emerging* from: something emerges into saying with which we can be in accord so as to respond, "to say after," which is what *nachsagen* quite straightforwardly means. Whether there is literal sound (cottonwood leaves rustling in the wind, a dog barking) or not, listening means gathered, heedful attentiveness. Anything else is in one ear and out the other. Words that say some-thing manifest saying in human speaking, opening a way for us to hear our own vibrating, intertwining relationship with all other things. Having heard, having gathered ourselves in hearkening attentiveness, having become attuned to our be-longing within the fourfold as mortal speakers of language, we speak in response.

Our saying-after the saying of things is no mere repetition. How could we, for example, repeat the rustling of a poplar or the hissing and crackling of a fire? And what, indeed would be the point merely repeating the words, no matter how powerful they might be, that we read in some text? Saying-after is, rather, our own peculiarly human responding to the call of the claim that saying has upon us (GA 12, 244, 251–4/WL 124–5, 130–4). Saying-after saying has its own dynamic energy, in that (1) it does not happen in isolation but in the mutual be-thinging or codetermining of us and things and (2) it participates in the way making of enowning and its saying, where the claim of saying first calls to us. It calls to us from out of our belonging with things, in timing-spacing-thinging, resonating in the thanc. It appeals to us in that it touches and moves us, calling us to recall and mindfully remember our belonging. In laying claim to us, saying calls for a response on our part; it calls for our saying-after saying. We do not need to rely entirely on Heidegger's German here. In English nouns the prefix *be-* can carry the sense of yielding or providing with whatever follows the prefix, so we can think this calling and claiming as a kind of be-speaking. In be-speaking us saying yields our speaking. It yields the words that we say-after saying, not in putting the words into our mouths but rather in calling forth our response. Belonging to timing-spacing-thinging and belonging therefore to saying, we are be-spoken, and we speak, saying some-thing.

No wonder Heidegger says that saying is the relation of all relations. The word most often used by Heidegger to say this relationality is *Verhältnis* and never *Relation*, though both words are quite correctly translated as "relation." The latter tends to imply a relation between two logically distinct or distinguishable entities or concepts. Since we are attempting to think be-ing, to think timing-spacing-thinging, it should be fairly clear why that kind of relationship is not applicable. *Verhältnis*, on the other hand, says the ways that things hold together (the root word, *halten*, is "to hold"). Heidegger puts into play some of the other words that also emerge from that root to open up the senses of "being held" carried by *Verhältnis*. "Language is, as world-moving saying, the relation of all relations [*Verhältnis aller Verhältnisse*]. It relates (*verhält*), maintains (*unterhält*), proffers and enriches the face-to-face encounter of the world's four regions, holds and keeps them (*hält und hütet sie*) in that it holds itself—Saying—in reserve (*an sich hält*)" (GA 12: 203/WL 107). World-moving saying (which is, in its own way, the same as enowning, way-making movement, timing-spacing-thinging, and be-ing) does not directly say *itself*. It holds itself in reserve, withdrawing and yielding to make ways for the interrelating and maintaining of the dynamic of thinging, holding and keeping things moving into the clearing, into the open, in the ongoing play of revealing and concealing, coming and going, moving-unto and removal-unto. This is much more than a connection between terms, concepts, objects, or beings; rather, it says the very movement that brings and holds together the fourfold so that gathering (thinging) takes place. This manifold holding is another way to say the insight that removal-unto (timing) is never merely dispersal, any more than moving-unto is mere coalescence into a being. There is, in the unique moments of timing-spacing, a stilling that is never a stopping. Only thus can showing ever be thought as *saying* that can, as the heart's core of language, make way for our thinking as coresponding and saying-after.

So it is that enowning-saying, this relation of all relations according to which we find our own belonging to timing-spacing-thinging, opens the ways for us to bring what is shown to language. Only thus does it become possible to think the first and other beginning. The manifold turnings in enowning yield the joinings of saying, with saying holding itself in reserve, echoing the withdrawing of be-ing as ab-ground, from which it is inseparable. *This* is what compels the reservedness that attunes an other beginning. Even the German word for reservedness—*Verhaltenheit*—is yet another of the words with "to hold" as its root and carries a double meaning of holding-in-reserve and comportment (how one "holds oneself," a with-holding holding-to-itself). In abandonment of being, be-ing is both withheld (reserved) and intimated. Likewise, the ringing of stillness within which saying resonates is withdrawn and held in reserve in the very dominance of the merely technical use of language. It does not simply appear, and yet it is at work, in that *all* language has saying as its heart's core. Heidegger speaks of withdrawing, whether of be-ing or of saying, as *sheltering* them (GA 12: 175/WL 81). What does this "sheltering" say? Note that that is less definite than saying "hidden" or "vanished." One of the phrases Heidegger also uses to say withdrawing is "hesitating self-refusal" (GA 65: 381/CP 266). This is another way to say the intimating that calls for thinking, the hinting that lets us become aware that there is more to the saying and showing (of both things and language) than is obvious. The intimation is that we can think this, *if* we approach it in a way that accords with how it holds sway, if we can let ourselves be attuned to be-ing's own reticent way making. We are attuned in our very heart's core, the heart-mind, the thanc: "*Reservedness* is the remembered awaiting of enowning because reservedness thoroughly attunes the inabiding" (GA 65: 69/CP 48). The dynamic of enowning-saying, always withdrawing and holding itself in reserve, attunes us to be *open* to the hints of the withdrawing that, in its silent saying, enables the gathering within which we belong. To turn with this ongoing turning in enowning requires that we corespond with our own reservedness in bringing its saying to language (GA 65: 35/CP 25).

In chapter 2 we also encountered another facet of the dynamic of saying and enowning, way-making movement. The way-making movement of enowning does not make way for thinging and its saying (showing) only so that we can hear it and respond in some dry, almost automatic or mechanical fashion. Way-making movement *moves us*. We are moved and touched by way-making movement because of our being thoroughly entwined with what arises in this movement. "To a thinking so inclined that reaches out sufficiently, the way is that by which we reach—which lets us reach—what reaches out for us by touching us, by being our concern" (GA 12: 186/WL 91). So way-making movement also enters into the thanc, sparking our inclination toward wonder, questioning, and deep contemplation that refrains from hasty grasping. If nothing else, placing our own deepest well of inclination, the thanc, into play with the saying-showing of enowning should counter our contemporary tendency to take both language and things for granted. We use them as if they belonged to us as property, seeing them only as objects on which to fasten concepts or demanding their constant presence and availability as standing reserve. In relation to things we are both "there" and "not there," not caring for things as things, not letting them show (and not-show) themselves, so that all too often we speak but do not say any-thing. In short, our own belonging to them is forgotten. Thus it is that we need a way (something that will let us reach) to where we already are, whether this "where we are" is called language or saying or the fourfold or be-ing or enowning or ab-ground (GA 12: 188/WL 93).

We need a way to where we already are, as those enabled to think in opening and tending the thanc.

To arrive at this juncture in thinking is to have begun to undergo a transformative experience of thinking that already starts to bring us into the region of dwelling. But we know already that dwelling is not only a matter of our relationship to language but of *staying with things*. We dwell insofar as we actually stay with things. In thinking with and after Heidegger we need to understand the relation of language, thinking, and dwelling, because if thinking is to reach its full enabling power, it will also unfold as dwelling. But thinking about that relationship, important though it is, is not yet dwelling as staying with things.

Dwelling means heeding and taking care of particular things, the things we make, the things we cultivate, and the things that arise on their own, if only we will hold back long enough to let them do so. Here we are called on to see another dimension of releasement toward things. Already, in first entering into the thinking of the first and other beginning, we begin to let go of grasping at "being." As thinking proceeds we also begin to let go of "beings," which *are* only insofar as they are grounded in some representation of being. The extremity of such representing emerges as enframing, reducing beings to standing reserve and reducing language to information used to control beings, us, everything. Releasement toward things here means letting go not only of clinging to the products of technology and to the limiting ways of calculative thinking. It also means releasing the impulse to have to control everything. Just as thinking the first beginning as such allows us to begin to see its concept of being as optional, so it is with the most extremely constraining representation of being, enframing. In seeing how enframing brings beings to presence as standing reserve we are able to understand enframing as another way of revealing beings. We are also able to think our relationship to it, heeding the danger it carries: the closing off of all other ways of revealing, thinking, and living. To think this closure already loosens its hold, but only if we *both* think it and let the thought sink deeply into our heart-mind, moving us in another direction, to be shifted from controlling beings to caring for things. And just as the first and other beginning arises as one, there is no sharp break here in saying that we care for things, whereas the long history unfolding from the Greeks was concerned with beings. The "past" is inside us, whether we think it or not. Heidegger's insight was that we need to genuinely think our historical relation to beings if we are to come into a dwelling relation with things. It is not as if there are (metaphysically) beings at one time and things

later. Be-ing does not somehow *actually* replace being. Be-ing says, primarily, ab-ground. It does so in joining with the other guidewords so that we
can begin to shift our thinking—or be shifted—from seeking constantly
present ground to being open to relationally dynamic grounding (ab-ground
as Ur-ground) (GA 65: 380/CP 265). Ultimately, all these words, words like
"ground" and even "be-ing" and "ab-ground," are, I would suggest, transitional and provisional.

In a segment of his dialogue with a Japanese acquaintance Heidegger
hints at something like this. The immediate context for this fragment of
conversation is the observation that many people did not understand the
intention of Heidegger's thinking of the history of Western thought, from
the opening question of the meaning of being onward. Listen carefully as
Heidegger (the "I" for "Inquirer") responds to some thoughts on his use
of language in that thinking.

> J: The fact that this dispute has not yet got onto the right track is owing . . .
>   in the main to the confusion that your ambiguous use of the word "Being"
>   has created.[2]
> I: You are right . . . [but] my own thinking attempt . . . knows with full clar-
>   ity the difference between "Being" as the "being of beings" and . . . as
>   "Being" in respect to its proper sense [which is, as *Contributions*—which
>   Heidegger had written over fifteen years before this dialogue—makes clear,
>   be-ing as enowning].
> J: Why did you not surrender the word "Being" immediately and resolutely
>   to the exclusive language of metaphysics? Why did you not at once give
>   its own name to what you were searching for . . . ?
> I: How is one to give a name to what he is still searching for? To assign the
>   naming word is, after all, what constitutes finding. (GA 12: 104–5/WL 20)

So it is that in thinking with Heidegger we still speak of being and be-ing.
But we also have the guidewords he gives us that resonate more deeply
into the matter, in consonance and dissonance, moving thinking at times
by tiny increments and at others by leaps and bounds, out from the limits
of metaphysical and techno-calculative thinking. I said very early that *none*
of these words, not even enowning (the "proper heading" of the touchstone
text, *Contributions to Philosophy*, is *From Enowning*), is the *last word* of this
thinking. Heidegger says that if we had such a word we would already have
arrived and found what we were seeking. We are obviously not at that point.
It is quite likely that there is no such point. In fact, I would suggest that

that is indeed the case. If what is ownmost to be-ing, to timing-spacing-thinging, to saying, is withdrawal from our thinking's grasp—their refusal to be captured in conceptual language—then why should we think that we will ever have "the last word" or the "best possible word"? If, as it must be, our responsive saying remains "forever relational," and the web of relations is through and through dynamic and ever-changing (in its timing-spacing: removal-unto and moving-unto), then the very notion of the last, best word to name this dynamic is somewhat absurd. If we thought we had such a thing, we would already be treating it as a concept and, in so doing, we would have twisted its saying power away from *saying*. What we are seeking here is not the last word but rather an ever self-refining joining within saying, a joining that resonates more and more deeply with our heart's core or heart-mind, the thanc, enabling us to *stay* with this saying (thinking) and only thus to also begin to stay with things (dwelling).

By now we should have begun to be aware of the sheer *energy* and power of language. It moves in us constantly, even when we are just daydreaming and—this is somewhat astonishing if we really think about it—when we are asleep. We babble and converse and make plans and do verbal battle even in our dreams. We are accustomed to think of language as our peculiarly human property and possession, but it seems that it is just as true—or even more so—to say that language possesses us. How we deal with that will make a huge difference in many, many ways. Within enframing we think we are using and controlling language, but in doing so we are only responding to the challenge (as a demand, a requirement, something that is assumed to be more or less compulsory) put upon us by enframing with all its dualistic presuppositions, as I outlined them in chapter 2. So we take up language as just another technique, as in advertising, media swaying of opinion, expert opinion, "studies have shown," political speeches, "disinformation." Or, in a variation on this theme, we take up language almost as a substance, another kind of being on call in standing reserve, to be used at will: language as information, language as commodity, "word processing."

So it is that the energy of language is ambiguous. On the one hand, we have the language that carries thinking, the guidewords and joinings, the language that calls to us and draws and points us in the direction of the withdrawing intimation of be-ing. On the other hand, there is the language of enframing. *All* of this language (and that includes even the crassest examples from popular culture) intimate in their own way the power and energy of language. Language can be referential, representational, objectifying, reifying, metaphysical, calculative, and playing out our illusion of being in control. Or it can be the language of thinking; it can be language arising from the thanc, language as coresponding, as saying-after, from the heart-mind, the thanc, language that responds to the saying of timing-spacing-thinging, as we stay with things, dwelling with and caring for them.

STAYING WITH THINGS

I think back to my once-only encounter with the mother woodchuck. If I had only known how to listen, I could have brought some of what she was saying to language. I know, in my heart's core, some of what she said, but in holding it so tightly to myself I never allowed it its freedom to emerge and *say* what needed to be shown. So even less was I able to say anything in response at that time. "Listen," she was saying, her paw resting lightly on my foot and her eyes on mine, "you and I are not so different after all. We are both of earth, in its dynamic power to bring us forth and draw us back into itself. Here I am, here you are, with my pups over there under the shed, and your growing vegetables over there not far from us. Here we are, *face to face*. How rare! Treasure this encounter, do not forget it, and one day you will understand what I am trying to say to you." To this day I stand in awe of the courage she showed in coming so quietly and without hesitation right up to what she might instinctively (and quite rightly) regard as one of the most dangerous animals on the planet. I can only think that she somehow knew that I was open to this, that something about me was momentarily different from the usual human closed-off arrogance toward the other animals.

Treasure this encounter, hold it, keep it safely in the thanc, because its wordless saying, emerging from the depths of great mystery gathered by what—according to our usual notions—is "only a groundhog," is endless, inexhaustible. This treasuring and keeping is the devoted recalling or re-membering that emerges in and as the thanc. Such keeping is a way of preserving and keeping safe what is to be thought (WHD 96–97/WCT 150–51). And it is this that enables us to begin to understand what it might mean to stay with things and to *hear* their saying.

The notion of *encounter*, of being face-to-face so as to hear and heed things, has a fairly significant place in what Heidegger says about staying with things. I have always thought it odd that he says so little about the other animals, given that one of the most clearly written and helpful passages on this matter concerns a face-to-face encounter with a tree. Let's see what we can learn from that and then return to the issue of our relationship with animals.

First, Heidegger brings this notion of the face-to-face encounter into the context of thinking and of thinging. He is quite clear that to "experience this face-to-face of things with one another . . . we must . . . first rid ourselves of the calculative frame of mind." Why? Because being face-to-face with one another "not only with respect to human beings but also with respect to the things of the world" originates in the joining gathering of the fourfold, that is (in terms of the language I have brought into play from "Building Dwelling Thinking" and *Contributions*), in timing-spacing-thinging. But as *face-to-face* encounter this is brought down-to-earth, so to speak, in order that "where this prevails, all things are open to one another in their self-concealment; thus one extends itself to the other, and all remain themselves" (GA 12: 199/WL 104). If we recall what Heidegger said of dwelling, that its staying with things is enacted in caring for and preserving them, it is clear that there is in that a strong connection with what he says here. Being face-to-face is a strong and decisive way to encounter and stay with something, and in reaching out and responding the things in this relation of mutual belonging are freed to "be" and say themselves, that is, to show what is uniquely their own. The most extended discussion by Heidegger of such an encounter plays out in *What Is Called Thinking?* on pages 40–41 (WHD 16–18). As I said, it involves a tree, specifically, a cherry tree in full bloom, with all the color and fragrance and motion that we can sense in such a tree. There we are, face-to-face with the tree, says Heidegger, and if we pause long enough to notice it, we realize that this encounter is not "one of those ideas buzzing about in our heads." Of course, we may speak scientifically about neurons and synapses and electrical impulses in the brain, all of which are undeniable. Nevertheless, there are some questions to ponder: "Does the tree stand in our consciousness, or does it stand on the meadow? Does the meadow lie in the soul, or does it spread out on the earth? Is the earth in our head? Or do we stand on the earth?" On the one hand, the answers to these questions may seem to be trivially obvious, even if lacking in scientific or psychological sophistication. Or we may think we are being asked to take a stand on the philosophical debate between realism, idealism, and antirealism. But it is just there, in both those ways of taking the questions, that we run up against the dominance of enframing, which makes us hesitate to simply *attend* to the tree as tree in its own place. Why do we not only hesitate to do that but, even if or when we do, then slink away and keep it to ourselves just as I did for so long with my encounter with the woodchuck? This "keeping to our-selves" is a far cry from the "keeping" of the thanc, which keeps something as a treasure to be pondered. The one keeping tends to set the encounter and its saying on the side, statically, with no way in or out; thinking sim-ply does not reach to the encounter in this case. The other keeping, the keeping of the heart-mind, reserves what is mysterious so as to recall and bring it into other encounters, other joinings in play; it becomes food for thought, it remains a matter for thinking. Why do we tend to set what is most meaningful aside in that rather craven way? It is not so much a per-sonal failing but a consequence of our having been educated and indoc-trinated to give over to others, to the experts (especially those who claim the backing of scientific method) the right to tell us what is or is not "valid experience," experience that is valuable and meaningful. The upshot is that "the thing that matters first and foremost is not to drop the tree in bloom, but for once to let it stand where it stands. . . . [But] to this day, thought has never let the tree stand where it stands" (WHD 18/WCT 41).

Heidegger is here attempting to provoke us to see how our tendency to grasp and cling, to reduce everything possible to an idea or a concept, pre-vents us not only from staying with things but from even genuinely encoun-tering them in the first place. I have already, more than once, spoken of releasement toward things as being applicable to letting go of this obsessive tendency to grasp at concepts. But there is more that we need to release if we hope to learn to dwell, to stay with things long enough to hear what they say to us. We must let go of our reliance on "the experts." I am not advocating some kind of reactive, wholesale rejection of science. Some of the best contemporary science is not nearly as rigid and limiting as what we are presented with in the popular science and "the studies" presented to us in the media. What I am suggesting is that we let go of (and at times this might mean forcefully tossing out) our passive, unquestioning accept-ance of anything that is presented to us as expert opinion, as authorita-tively reflecting what "they say." As early as *Being and Time* Heidegger let us know some of the ways that we are shaped, constrained, closed in, and closed off by this "they." We can become and in fact ordinarily are so molded that our very "self" becomes, effectively, a "they-self." The they-self does not think, not in the sense in which we are now using that word. The they-self goes rather mindlessly about its business, business that has also been laid out for it by "them." Stop and think: How much of what we call our beliefs, our ideas, our values, and our lifestyle comes not only from the

Western philosophical tradition but also from Wall Street, Madison Avenue, and Washington, D.C.? In its interactions with things the they-self is incapable of face-to-face encounter, of hearing what things say, of heeding any intimations of timing-spacing-thinging. The they-self does not even have the time to stop and pause long enough to wonder about this, being herded about and channeled within the framework of public time, clock time (GA 2: 168–73, 221–26, 259, 544–62/BT 163–68, 210–14, 239–40, 464–78). The they-self is incapable of dwelling unless it can undergo radical transformation.[3]

Dwelling with things, as nurturing those that grow and constructing those that do not, first of all requires *encountering* them face-to-face, which is to say, encountering them where and as they arise and come forth to meet us in their own place (timing-spacing). In the dynamic, multiplicitously sensuous play of relationships, face-to-face does not mean standing there and staring at something. I have learned some important things while sitting at the base of an ancient longleaf pine. Face-to face can be back-to-back. The most important thing at this stage is to realize that if we bring to this encounter a heavy load of presuppositions from whatever source, the *intimacy* that is indicated by the phrase "face-to-face" will be ruled out before it even has a chance to arise. We might say, "So what?" In terms of our usual habit patterns the very notion of "intimacy" with a tree or with a wild animal (such as, say, a woodchuck) or with a rock seems odd at best or perhaps even absurd (i.e., meaningless). But if we have thought along with Heidegger to this point and find that any of what has been said of timing-spacing-thinging rings true or resonates within our heart-mind, then we can also see that not only is this intimacy meaningful, it is necessary. Why?

Because, whether we take the time to think it or not, we are already thoroughly entwined in the relational dynamic of thinging. Intimately encountering particular things yields us a way of access to this dynamic's showing, its saying, in such a way that we *can* respond in accord with it from the thanc. We can think "thinging" in general terms, to a point. But that is not yet dwelling. Dwelling requires owning our enowned belonging to saying through *listening* to it. We listen to saying first of all through staying with things and then also to thoughtfully listening to what comes to language from within this heedful staying. Dwelling, staying, listening, thinking: all belong together, saying and enacting the same. Our own thinking, our language, and even our very lives emerge from within the matrix of timing-spacing-thinging, which is the ongoing gathering and dif-fering of things.

I should perhaps point out that my use of the word *thing* seems to be a bit broader than Heidegger's. I am not so sure if it is or not, since there is some ambiguity at work here. It would, I suppose, be rather surprising if that were not the case, given that "thing" is one of the most all-encompassing terms in our language. At this point I will simply say that, insofar as "thing" means gathering and arising within and as timing-spacing-thinging, I see no good reason to limit the application of the word to so-called inert things, perhaps including the plants among them. (Of course, plants are themselves anything but inert, as any gardener can quite plainly see every day.) So, yes, in one sense I think my use of "thing" is broader than Heidegger's. I say that animals, *in this sense* (arising in and as the gathering of timing-spacing-thinging), are things. *And so are we.* However, this does not mean that I am now somehow "reducing" us—or the other animals—to the status of so-called mere things. Within the context of this path of thinking, attempting to think for the sake of dwelling, it is the very notion of "mere thing" that lacks meaning. *Any* thing, gathering as it does the whole of its relations to other things in timing-spacing, is profoundly rich and mysterious. To stay with our fellow things, to push this thought a bit harder, is to be open to the entire matrix of thinging, of saying, and of meaning. It is to be open to the entire matrix, as it comes to meet us in *this* always-unique thing, whether it be oak, rufous-sided towhee, staple-gun, or pan of brownies.

We return to the cherry tree in full bloom. I do not need to imagine Heidegger's cherry tree, having one nearby on the hilltop. Unlike most of the other wild cherry trees in the area, this one is apparently one that escaped from cultivation decades ago, perhaps carried as a seed in the belly of a grackle or robin. In late March and early April it comes into bloom. This is a very old tree, standing at least forty feet tall and nearly as round, with thick, lush foliage. Before the leaves burst out from the branches, however, the blooms arrive. When they open the tree is almost completely covered with glistening, pure white flowers. It is stunningly beautiful, especially against the blue of the sky and with the deep rose of the redbud trees blooming below. I know that if I were driving on two or three of the roads nearby I would be able to see the tree from up to a mile and a half away. This blooming cherry has stood there through thousands of sunrises and borne the weight of many snowfalls. It has weathered ferocious thunderstorms and accepted the most gentle of rains. Hundreds of thousands of birds have roosted on its limbs, made their nests within its sheltering branches and foliage, and eaten its fruit. I stand under this cherry tree, listening to the bees working the blossoms and the leaves moving against one another in response to the slight breeze. I wonder, What does this tree know that I do not know? I hope it is clear that in putting the question that way I am not anthropomorphizing the tree. On the contrary, it is the very nonhuman uniqueness of this tree that speaks to me, that has *something to say* to me.

It is this uniqueness of things that should also avert some misunderstanding of what I say about our thingliness. Our uniqueness includes having the kind of brain that allows us to think discursively and develop our elaborate linguistic responses to and ways of saying what comes forth to meet us in the play of timing-spacing-thinging. But again, the only reason we can do so is because there is something coming forth, because we are always already within this matrix of emerging showing forth that we may call saying, or enowning, or be-ing, or names as yet unthought. And, as timing-spacing is always thinging, most of that sense, that saying, is carried to us by way of our encounters with things, which also carry the intimation of the withdrawing of timing-spacing-thinging itself.

What I say here about us also somewhat indirectly supplies a response to the question concerning animals that I have tentatively raised more than once. If we are things, so too are the other animals. But, again like us (and all things), they have their own unique ways of gathering the relations. There is never homogenization or mere identity but, rather, dif-fering into what is unique in each dynamic opening-extending moment. As Heidegger puts it, be-ing's holding-sway (which, as it is not a being, is inseparable from things) is "unique and once only" (GA 65: 385/CP 269). Thus there is both kinship and distinct difference between animal and animal, and between us and the other animals, and between all things. As Abram puts it, drawing on his extended dwelling with oral tribal cultures and his attempt to understand their ways of thinking about animals, "The other forms of existence we encounter—whether ants, or willow trees, or clouds— are never absolutely alien to us. Despite the obvious differences in shape, and ability, and style of being, they remain at least distantly familiar, even familial. It is, paradoxically, this perceived kinship or consanguinity that renders the differences so obvious, so eerily potent."[4] In their very difference- within-kinship what they have to say to us is all the more telling.

I said that sense is carried to us by way of our encounters with things. I want to follow that thought further, as well, and see where it leads. Without things there could be no meaning, no saying, no language. That is as much as to say, without our encounters with things we would not be human in any recognizable sense. Here my thinking converges rather closely with that of Merleau-Ponty and Abram (who is inspired by Merleau-Ponty as well as by fireflies, spiders, and other interesting characters). Merleau-Ponty puts forward the thought that flesh (his word for the dynamic, intertwining relationality that I call timing-spacing-thinging) is not chaotic meaninglessness; rather, "there is a strict ideality in experiences that are experiences of the flesh: the moments of the sonata, the fragments of the luminous field, adhere to one another with a cohesion without concept, which is of the same type as the cohesion of the parts of my body, or the cohesion of my body with the world."[5] Merleau-Ponty thinks this cohesion dynamically, in terms of a vast field of intertwinings that is, just as I have said, the matrix from within which the very possibility of meaning arises, such that "language is everything, since . . . it is the very voice of the things, the waves, and the forests."[6] Change "language" to "saying," and he says precisely what Heidegger says and what I say along with them. Thus it is that Abram also refers, throughout *The Spell of the Sensuous*, to the earth itself as animate, as en-spirited, taking a cue from the indigenous cultures who dwell as participants within a landscape that is never "definitively void of expressive resonance and power: any movement may be a gesture, any sound may be a voice, a meaningful utterance."[7]

This brings us back, then, to these questions. How do we listen to things? How do we stay with them so as to free them while nurturing or caring for them? Assuming that we are first able to sufficiently release or let go of some of the most compelling hindrances such as techno-calculative thinking and perceived lack of time, we can each respond only from within our own situation. There can be no method of dwelling any more than there can be a method that constrains and shapes thinking. So I will speak from out of one thing I do rather well—cook—and perhaps it will also say something that points more generally to ways of dwelling, of staying with things.

COOKING-DWELLING-THINKING

There it is, near the end of the raised bed: a perfect sandwich tomato, a dead-ripe 'Dixie Golden Giant'. Glancing a few feet on down the line, I see a good prospect for tomorrow or the next day in an unusually large 'Red Rose'. That variety may not have quite the intensity of flavor of its parent, 'Brandywine', but it will give us quite a few more sandwiches, sauces, and soups. The June bugs apparently approve of it. Three of them are feasting with their heads down in a crack in one of the tomatoes. They have to eat, too. As I carefully pull the big yellow tomato from its vine, the pungent odor of the foliage clings to my hands and reaches my nose. Into the house we go, with the fruit warm from the sun, heavy in my hands. This one is big: a two-sandwich tomato. My sister and I have long agreed on what is needed for the perfect tomato sandwich: two slices of commercial white bread (as fresh as can be had), a thick slab of just-picked fully ripe tomato from a fruity-flavored variety (there is a just-right range of balance between sweetness and acidity), and plenty of good mayonnaise. That's it. Anything else would just be so much gastronomic clutter. Furthermore, a perfect tomato sandwich will be eaten standing over the sink, with the juice (not too much, mind you, as this should be a meaty tomato) running down your chin. Aaaaaaah . . . that is *so* good.

I hope we all have several things that taste that good to us. Of all the things with which we interact on a day-to-day basis, the things we eat are both the most essential (along with water, of course) and the most *intimate*. More obviously than most things, they enter directly and decisively into us. Reciprocally, how we treat them decisively alters them. Growing, cooking, and eating constitute as clear and easily accessible an array of possibilities for considering the relational dynamic of dwelling as I can think of.

Why do I make the tomato sandwich in just that way? Because the flavor is held in my memory in all its nuanced richness. It is, to use the language of thinking, held in safekeeping in my embodied memory, within my heartmind, the thanc. In fact, physiologists and psychologists have observed that for most people experiences of odors and taste, which are so closely entwined as to be inseparable, are the most precisely remembered of all our experiences. Taste and smell are the most visceral and unforgettable of the senses, but what we experience in that way is often very difficult to bring to language. We have many precise words and, in some cases, even mathematical descriptions for shapes, colors, musical tones, loudness, and different textures, but, other than a few rather general terms (salty, sweet, sour, bitter, along with strong and weak), when it comes to taste and odors we usually have to say something like "that smells like skunk" or "this tastes a little like a tangerine."[8]

Not only is the saying of the kinds of things that we call by the name of food wordless on their own part, but our own saying in response may be more silent than not the more fully we co-respond to what comes to meet us in the encounter. (The social character of much of our eating takes the matter in another direction that is beyond the scope of this chapter.) The "aaaaaah" as saying-after the saying of the perfect tomato sandwich is not the least bit vague or imprecise, though it is nothing like a concept. It grasps nothing while saying manifold timing-spacing-thinging, gathering the relational dynamic of the tomato, bread, and mayonnaise (and the whole of the net of Indra gathered in them) along with my embodied keeping in heartmind the cumulative taste-memory of a long line of tomato sandwiches. From there, from the thanc, emerges the grateful thought of responding to the thinging of this thing, this tomato, by placing it in my hands and eating it with the most heartfelt mindfulness I can hold.

This gratitude is enhanced by my awareness of all it took to bring me this tomato: every detail of what Heidegger calls the fourfold had to connect "just so." Earth, he says, bears up and bears forth. This is not only Earth in the large, planetary sense but this very patch of soil, with its millions of microbes, its earthworms and humus and minerals. This earth must be tended and nurtured if it is to say "tomatoes" rather than "thistles." So it is that I have returned compost to the soil, mulched the plants, and pulled the unwanted competitors, the "weeds." The sky is the most fickle of the four. Nine of the last ten years in eastern Tennessee have been drought years. This year we have had plenty of rain. Everything is green and lush except for the tomato vines. The unusual wetness of the summer has brought along a new blight, some fungus, apparently. So each tomato that makes it to ripeness without rotting from the core is a gift and a blessing. What about the divinities? After more than thirty years of gardening I find that each tomato, each peach, each ear of corn is—if I pause long enough to really look, to listen, to think—amazing and mysterious. I know quite a bit about the art, craft, and even science of gardening (e.g., the nutrients that are needed for the different kinds of crops). But even so, this unique yellow beefsteak tomato, soon to be a sandwich, intimates the deep mystery of timing-spacing-thinging. It says and shows itself while hinting at the deep workings of what is, ultimately, inconceivable. Inconceivable but by no means nothing. If I let go of my tendency to grasp things conceptually, I can just begin to in-grasp what the tomato says. This tomato is soon to come to an end as a tomato as I swallow the last bite of my sandwich. Just so, at some as yet unknown, unique moment, this life of mine will reach its culmination. We, all things of whatever kind, are together in the play of moving-unto and removal-unto, spatializing and temporalizing. In my heart's core, the thanc, I am moved to the deepest, gut-level realization: this unique and unrepeatable moment of timing-spacing-thinging is to be treasured.

A perfectly delicious tomato sandwich is a simple, ordinary thing, and yet it carries within itself the deep and complex mystery of be-ing, of timing-spacing-thinging, just as I discussed by way of the loaf of bread in chapter 3. (By the way, semolina-sesame bread makes a very good tomato sandwich, too.) I participate in its thinging, its saying, as it evokes my saying response. This is quite obviously not only a matter of oral or written language. "The hand's gestures run everywhere through language.... Every motion of the hand in every one of its works carries itself through the element of thinking, every bearing of the hand bears itself in that element. All the work of the hand is rooted in thinking" (WHD 51/WCT 16). Dwelling is the joining of thinking and other activities that encounter and correspondingly stay with things. Listening for their saying, in mindful awareness of our mutual belonging, we say-after what shows itself in this unique thing through our saying response: speech, silence, the work of the hands, and at times tasting, chewing, and swallowing.

Dwelling does not take place in the framework of the old active-passive dichotomy. The saying of things is unique in each moment, and so is our saying response. It is unique but not fixed in the first moment of encounter, and it is not predetermined as the encounter unfolds. In the staying with things that constitutes dwelling we care for and preserve what is each thing's own. But what is ownmost to any thing is hardly ever so puny as to admit only one possible response. This 'Red Rose' tomato could have been the heart of an outstanding tomato sandwich if I had not just eaten that sandwich made with the 'Dixie Golden Giant'. Here we have a basket with more tomatoes, though—several 'Burgundy Travelers' and a couple of 'Royal Hillbilly' beefsteak tomatoes. They tell me that, depending on the relational context (Is there some basil ready to pick? How are the eggplants doing today? What about the zucchini? I pulled and braided the garlic the other day), I could let them recite their poetry in at least three ways. I could slice them onto a plate, drizzle them with good olive oil, and sprinkle them with coarsely chopped basil. Salad, just like that. Or I could peel, chop, and toss them, just as they are, along with some basil, rosemary, and red pepper flakes, with some hot pasta and olive oil. However, if the eggplants and zucchini are loudly calling me and the bell peppers have something to say as well, I could make some slow-simmered marinara sauce.

It is no doubt obvious that to proceed this way, along the paths of the art and craft of cooking, requires some know-how, acquired through experience, embodied and kept in the heart-mind. This know-how, if it is to play out in dwelling, is not reducible to technique. "Cooking must express taste, not technique, because technique alone does not communicate anything.

To study it otherwise than as a function of taste is an arid academic exercise; it is like mastering the grammar of a language in which you have nothing to say."[9] Too much emphasis on technique means an emphasis on having to be in control. As a passionate cook I will assert flatly that relating to food from within the confines of enframing sets us up to settle for inferior, second-rate food. The clearest example of this is probably the chain fast food restaurants. Why are they so successful? It is because—to be fair—what they produce seems to taste quite good to many people. But that would not be enough. The other significant and perhaps most important factor is that there are no surprises. A Big Mac is a Big Mac is a Big Mac in Atlanta or Anchorage. The rule is this: put out a consistent—and that means *uniform*—product. One Big Mac is interchangeable with any other Big Mac. As for that tasty hamburger flavor, it originates, for many of the fast food chains, in a chemical flavoring concocted in a factory somewhere in New Jersey and sprayed onto what would otherwise be a relatively tasteless ground beef patty.[10] This is a far cry from venison carefully grilled at home and placed on a homemade bun, along with—here it comes again—a nice juicy slice from one of those tomatoes. On the one hand, you have technique, with the burger and all the ingredients on call in the standing reserve of uniformly available foodstuffs.[11] On the other hand, there is craft and perhaps even art: staying with things, listening and responding to what they say to us in the joining of thinking and dwelling. And instead of insisting on uniformity, we are open to being surprised by the outcome, perhaps adding to what is kept and treasured in the thanc.[12]

Obviously, I am speaking from the viewpoint of someone who enjoys cooking, is very skilled at it, and loves to eat. That description is not going to pertain to everyone, nor should it, necessarily. But our relation to the food we eat is one of the most vivid and easily understood paths by which to gain a basic understanding of what dwelling means. I hope it is also clear that in opening a way into the matter in this manner I do not mean to suggest that cooking as dwelling requires years of experience and a vast repertoire of exotic gourmet recipes. One just-picked fully ripe fig *speaks for itself.*

Recall that in "Memorial Address" Heidegger said that everyone can think if they only dwell on what deeply concerns them—what moves their heart-mind—here and now. The dwelling in thinking is none other than the thinking in dwelling. And so it is that just as we can all begin to think, just so, we can learn to dwell with things. And since thinking and dwelling belong together, they are opened to us as possibilities in the same way:

through *releasement toward things* and *openness to mystery*. The mystery we encounter is at bottom the same for both thinking and dwelling: the dynamic of showing and withdrawing in timing-spacing-thinging. This mystery is intimated in each genuine encounter with things. And not only that. Think back to one of the most significant things Heidegger said in *Being and Time*, something so important that he italicized the entire sentence: *Dasein is its disclosedness*. This disclosedness, in that text, is the structure of Dasein as the t/here, the opening for the manifestations of the relational dynamic of the beings of Dasein's world. In the thinking of the first and other beginning that is just the opening play, hinting at the mystery yet to be encountered in the attempt to think be-ing. One of the things we need to carry forward from *Being and Time* is the hint that *opening* (thought in both its senses, as noun and verb) will be of central significance for as far as thinking extends. We release the things that hinder thinking and dwelling, releasing them into opening, insofar as we are able to stay open.

We can think and dwell because of our belonging, with things, to and within timing-spacing-thinging. Think of all the times I have already used the phrase "dynamic relationality." One thing that falls by the wayside is any kind of dualism. This relationality is much more complex than that. It is echoed in the guidewords that "say the same" not as an identity relation, but as a belonging-together. Begin thinking with any of them, and the others begin to move into the clearing, bringing something that also needs to be said. Whether we begin with timing-spacing-thinging, or things, or be-ing, we hear also in those words the intimation of any or all of the others: saying as the "relation of all relations," the play of arising and withdrawing, saying as the ringing of stillness, thinking as welling up from the thanc, holding the mystery and its hinting in the heart-mind, and our belonging intimately within the entire relational dynamic, hearing its voice in each thing that we can encounter face-to-face. The one facet that has not yet received its due share of our thinking is opening. It has been mentioned in several ways, from time to time, all the way from that statement in *Being and Time*, to the way that timing-spacing is an opening-extending that makes way for (and is) thinging, and my ongoing emphasis on the importance of our remaining open to mystery.

So it is clear that being open is of central significance. But what, really, does this mean? Clearly, it means much more than what is encompassed under the rubric "open-minded." Insofar as it goes with releasement toward things, it implies also a willingness to let go of habit-patterns, even when what might come in their stead is not yet clear. It thus includes being able to hold a question without grasping at an answer. The hardest thing to release, however, and the hardest habit-pattern to break is the tendency to reify ideas and things. In the thinking of the first and other beginning, moving as it does from the first inkling of abandonment of being, through the recognition of the creative invention of the ontological difference, to the thinking of be-ing as ab-ground, it is clear enough *in thinking* that reification is the style of metaphysical thinking and that it blocks the attempt to think be-ing. That is clear, but to hold to it, to release the tendency to reify, is not easy even for thinking. Thinking and dwelling, however, belong together. It is even more difficult to hold back from reifying things than from reifying our ideas. Listening deeply to the saying of things, however, we will find nothing that stands still long enough for reification. Things are simply not beings. Heidegger says it quite directly, even bluntly: "If we stop for a moment and attempt, directly and precisely and without subterfuge, to represent in our minds what the terms 'being' and 'to be' state, we find that *such an examination has nothing to hold* onto. . . . We notice at once . . . that being is not attached to the mountain somewhere, or stuck to the house, or hanging from the tree. . . . We notice, thus, the problematic that is designated with 'being'" (WHD 137/WCT 225–26). "Being," as we have already seen, is a concept, with nothing to hold on to in the play of timing-spacing-thinging. Just after that comment, Heidegger then suggests that to move the thinking forward from that insight, we "give our heart and mind," that is, the thanc, to particular "beings," to things.[13] Again, thinking converges with dwelling. The thanc moves us powerfully, perhaps, but the power of the habitual tendency to reify things is as persistent as the Terminator. One of the most powerful scenes of that popular film was the one in which the hand, which was now all that was left of the Terminator, crawls forward, as mindlessly intent as the machine it was, to carry out its programmed mission. In a sense, we all carry in us at least two thousand years of programming. In both English and German, words for dwelling (such as habitation or *Wohmung*) and for habituation (*gewöhmen*) are linguistically close kin. Ordinarily, we dwell in accord with our habitual ways, with what we are accustomed to, the usual and the typical. And nothing is more usual and ordinary for us than to think of things as beings. *Even more difficult is to let go of the habit of reifying ourselves.* And yet that habitual tendency is the one most likely of all to stand in the way of thinking and dwelling. But how, we might well ask, could we let go of that? And why, indeed, should we? Isn't that going just a little too far? Thus speaks the reified self. To seriously think that issue, which is needed before even wondering how such a shift could take place, is what is needed at this time. If we are to be able to let go of the reified self and remain open to mystery in a deep and abiding way, it will not be a matter of an act of will (more control, there) or a change of attitude. It must grow from out of the overall shifting that arises in and as thinking, from the deepest motive energy of the thanc. It is the task of chapter 5 to inquire into how (and why) that becomes possible.

5

The Radiant Emptiness
of Be-ing

If we stop for a moment and attempt, directly and precisely and without subterfuge, to represent in our minds what the terms "being" and "to be" state, we find that *such an examination has nothing to hold* onto. (WHD 137/WCT 225)

Contrary to our habitual ways of thinking, shaped and reinforced by over two thousand years of metaphysical thinking, the thinking of the first and other beginning comes to this realization: reification is a conceptual move with no actual basis. We reify our ideas, we reify things, we reify ourselves. Nothing is more usual and ordinary for us than to think of all of these as, in one way or another, beings. Yet, as Heidegger says, the farther we go in the questioning attempt to think be-ing, the more we find that this usual, habitual thought-pattern has nothing to hold on to. Therefore, it would seem fairly clear that clinging to reification will hinder further thinking. But it will also hinder any attempt to explore the possibility of dwelling, of staying with things. There is simply no basis to consider a thing, as the gathering of the fourfold in dynamically relational timing-spacing, in terms of its being reducible to "a being." The play of the dynamic relationality of enowning is not reducible to something so flat, unless it is just another abstraction. We do not, it is to be hoped, *dwell* with abstractions but rather with things. But who or what is this "we," this "I," that dwells and that raises this question of dwelling? That is a question that emerges from the reflections in chapter 4, on which we will ponder over the course of this chapter and the next.

Before coming to a clear sense of the meaning of this "I" that wants to think and wants to be open to the possibility of dwelling with things, it is necessary to directly confront the crucial place and full ramifications of the nonreifiability of be-ing, of things, and even—I say "even" because this is the most difficult of all for us to accept—of us. As Heidegger puts it in one place, "In the question 'who are we?' is lodged the question of *whether we are*. Both questions are inseparable" (GA 65: 51/CP 36). We begin, of course, with the full awareness that this is not a question of the sort asked by Descartes 350 years ago. We do not wonder, as a matter of doubt (whether it be merely strategic doubt, as with him, or genuine), if anything of our experience of ourselves is "real" or "certain." Here, in the thinking of the first and other beginning, the question of *whether we are* is a question that first begins to fully confront our own place in the thinking awareness of abandonment of being, in the enactment of enframing, and in opening the possibilities of an other way, other *ways*, of thinking and dwelling. It asks us to confront and think through our own situation in regard to being, be-ing, beings, things.

The question concerning *whether we are* asks, *Are* we? Taken strictly, that asks whether I am a being. In the first four chapters of this book we already have several indications that the answer to that question is a simple "no." However, because that response is—no matter how unavoidable—rather startling, let me begin by briefly recapitulating the things that have pointed in that direction in the first three chapters.

Chapter 1. In the historical thinking of the first and other beginning we learn that "being" emerged out of the ancient Greeks' thought in response to their wonder at the arising into presence of beings. Wondering, they asked, What is this, the beingness of beings? Asking in that way, they sought for what is most general to all beings—their presence—and first differentiated this presence—the being *of* beings—from those beings. Not noticing the creative nature of these thoughts but, rather, taking them as if they were a discovery of the way things actually are, the question of the arising (*physis*) of beings does not come in for further scrutiny. Taking being as the ground of beings, the way is now clear for interpreting that ground and the beings it grounds in various ways, yielding the basis for determining ways of relating to those beings as interpreted. Thus we have the unfolding of the history of Western metaphysics; in all its permutations it holds to the notion of being as grounding presence, and of beings as what is present, without explicitly *thinking* any of this as such. In the contemporary period this emphasis on what is general and common to all beings culminates in the rule of being as enframing, with the peculiar consequence that beings begin to lose any remaining standing as unique beings or even as objects and are taken as interchangeable units of production in a standing reserve of such "beings." In the thinking of being as enframing we encounter a hint of what Heidegger calls "abandonment of being." The functional idea of "beings grounded on being" begins to lose its hold. The abandonment of beings by being, however, echoes the originary abandonment of beings by be-ing in the first beginning. And this, in turn (i.e., in turn for thinking, though in be-ing it is not this linear), echoes be-ing's originary holding sway in always withdrawing in opening ways for the arising of beings. To put that thought in other words, *be-ing "is" ab-ground*, the refusal and staying away of ground. What that says, however, is that be-ing is not being; that is, be-ing is not in any way thinkable in terms of beings. In fact, be-ing is nothing at all, says Heidegger, either in itself or as grasped by some subject (GA 65: 255, 484/CP 180, 341). Be-ing is therefore not reifiable. But since being serves to ground and determine the nature of beings, if there is no being, how can there be "beings"? This is the first indication that the reification of beings is a matter that is seriously questionable.

Chapter 2 aims to help us reflect on how to think be-ing. That is, how can thinking move in ab-ground? What sort of language is called for in this thinking? One of Heidegger's suggestions is that we approach the matter through releasement toward things and openness to mystery. To even begin it is necessary to release old assumptions about the nature of thinking, the thinker, and about the language that carries thinking, *because those assumptions arose together with the history of being.* They fostered the unquestioning acceptance of the ontological difference and were further reinforced in the ongoing attempt to enact and refine the various metaphysical interpretations of being. From the front end (method) through the middle (concepts and representations) to the end (theory and system), metaphysical language cannot say be-ing, which is ungraspable. However, this inconceivability of be-ing does not mean that be-ing is not thinkable at all. Being is intimated in the arising of beings (in full awareness of its ambiguity, that word remains, at least for a while, in use here), in their showing forth, which is, as Heidegger puts it, their *saying*. But this saying of beings is at once the heart's core of language that, if it is actually language, *says* or shows something. This gives us a strong clue that what we need, if we are to think nonreifiable be-ing, is a changed understanding of language. I suggested that we begin by taking the guidewords Heidegger gives us as joinings, akin to the dynamically mirroring jewels of Indra's net. None is definitive, none is the last word; they only say be-ing as ab-ground in their ambiguously resonating interplay. If we are to think the nonreifiability of be-ing, the nature of the matter itself requires us to engage it with non-reifying language.

Chapter 3. If beings are not "beings," how are we to think them? If be-ing is not "being" and thus is not "ground," then what is its meaning? Its "nothingness" is not vacuous, void meaninglessness. Ab-ground, says Heidegger, is Ur-ground: primal ground*ing*. Grounding for what? Things. Heidegger takes up this most ordinary of words and gives us a way to begin to think some of the most difficult questions in all of philosophy in a much more concrete way. Things are not beings but *thinging*, which is the gathering of the relevant web of relationships (other things, supporting and self-enclosing earth, sky with its blessings and its vagaries, the intimations of deep mystery, and us). This gathering is not randomly chaotic but takes place as the moving-unto one another of what is drawn together (spacing), while the dispersals of removal-unto (timing) make a way for the gathering to emerge as a thing in all its momentary uniqueness of be-ing. The entire discussion of timing-spacing-thinging strongly emphasizes the non-reifiability of the matter for thinking; timing-spacing is, as the title of section 242 of *Contributions* says, ab-ground. That discussion also hinted that we would, at some point, be called on to confront the question of our own reifiability. As the mortals in Heidegger's poetic fourfold we are implicated in timing-spacing-thinging. And the further we take this thought, from its first emerging in chapter 3 all the way through chapter 4, the harder it is to evade the insight that the description of timing-spacing-thinging includes us as more than just a bit player. We, too, emerge in the dynamic relationality of timing-spacing-thinging. This hints that, along with the demise of old assumptions about being and beings, comes the deep questionability of our long-cherished, strikingly dualistic notions about ourselves: rational animal, subject in a world of objects, and mind in a body, distinct from all other bodies.

The key to beginning to think this through more clearly is to take up the "nothingness" that pertains to be-ing as the matter to be held in question.

BE-ING: NOTHING

Be-ing is nothing. That certainly pulls away from reification. However, if left there, it pulls too far to the other extreme. Heidegger also says of be-ing that "we cannot equate it with the nothing" (GA 65: 286/CP 201). What does this mean? It means, in the first place, that our tendency to grasp at a reifying concept is so strong that when we first attempt to think be-ing as ab-ground there seem to be only two options: nihilistic nothingness (utter meaninglessness) or the reification of nothingness. The first of the two is both so abstract and so dire that one tends not to linger there unless as some pointless academic exercise. Heidegger is quite clear that to take this "nothing" of be-ing as simply nihilating is to misconstrue it (GA 65: 246/CP 174). Be-ing, after all, is another way to say timing-spacing-thinging, which is obviously not "nothing at all" in the flatly nihilistic sense. So it is not at all difficult to dispose of the nihilistic sense of the "nothing" that pertains to be-ing. The other sense is quite another matter.

Be-ing is ab-ground, the staying away of ground. Because be-ing is not a being, neither can ab-ground be conceived as a being. But because (1) the tendency to reify is so strong and very nearly compelling and (2) conceptualizing is quite often in play with another creative impulse, imagining, it is difficult to refrain from reifying nothingness, to hold back from conceiving it as "*the* nothing." For one thing, the word *Abgrund* can be translated as "abyss" as well as "ab-ground." I have on occasion translated it that way myself. But then there is a tendency to imagine a huge, gaping abyss, which is, of course, not really "nothing" but "something." It is a void sort of being, rather akin, in its own way, to the modern idea of parametric space. But we have already thought through the derivative character of that notion, both in terms of Dasein's temporality and in terms of timing-spacing-thinging. There is no way to reduce Ur-grounding timing-spacing ab-ground (i.e., be-ing) to being, abyss, time, or space, all of which have been shown to be derivative abstractions. What then? How are we to *think* this "nothing" that pertains to be-ing? We know that it says, in yet another way, that "be-ing is *not* a being," but what more does it tell us?

This "nothing" must be brought into play in the joining of all the guidewords that belong together as partial ways of saying be-ing. That means, of course, that there will be several ways into and through the matter in question. As a reminder of how joinings of guidewords work to unfold the meanings of the matter and as a way of looking for hints about how to begin to think this "nothing," I will take up several of the most pertinent guidewords and lay out how they relate to the others. The basis for this is that we already know that the guidewords belong together as "saying the same"; that is, they belong together as ways of entering into the thinking of what Heidegger calls the "simple onefold." Any one of a great many passages of *Contributions* pulls several of these guidewords into relation with one another. For instance, we have "Be-ing: enowning, nihilating in the counter-resonance. . . . Ab-ground: as the time-space of the strife . . . of earth and world." All the way back to the early "The Origin of the Work of Art" (1935–36), the mention of the strife of earth and world intimates thinging, which is there described as the play of the concealing pull of "earth" and the revealing of "world," the humanly articulated world of meanings. This gives us a start on articulating the first guideword (be-ing) in terms of the others, in how it indicates some of the ways that be-ing is in play with them.

Be-ing. Be-ing is said, just above and elsewhere, as both enowning and ab-ground. As enowning, be-ing says the gathering that, in thinging, brings things into their own, where they can appear and say (show) themselves. As ab-ground, be-ing is said to be *nihilating* in the counterresonance of time-space, that is, in the removal-unto and moving-unto of timing-spacing, in the dynamic of the no-longer and the not-yet that opens and extends the clearing of thinging. In the ongoing changing that is thinging, removal-unto "nihilates" what was, which never actually "is" (as something simply present). But as removal-unto is also gathering, in play with moving-unto, this displacing never just disperses but also stills and stays, revealing things while concealing thinging (be-ing, enowning, timing-spacing). In yet another way this says ab-ground.

Ab-ground. The tendency to reify is so persistent that even the thought of ab-ground can fall victim to it if we begin to think of it as merely a vacuous, gaping abyss. On the contrary, "Abground [holds sway] as time-space . . . the site for the moment . . . the strife of earth and world" (GA 65: 29/CP 21). This affirms what I just said in gathering some thoughts on be-ing. We can now say that somewhat differently: the staying away of any constantly present ground (being) allows the play of removal-unto and moving-unto as well as of revealing and concealing (withdrawing). Notice the twofold dynamic. Revealing/concealing and removal-unto/moving-unto are by no means identical, yet they are in play in the same timing-spacing-thinging. Removal-unto enacts (gathers) the withdrawal and concealing of what passes away in the gathering that is moving-unto. In this resonating gathering timing-spacing itself (be-ing, enowning), as ab-ground, stays away, withdraws, is concealed. It is not concealed in some black hole of utter voidness, however, but is intimated in the very play of removal-unto and moving-unto. Thus Heidegger says that ab-ground is the *hesitant* refusal of ground and that as such it is Ur-ground, primal grounding, "self-sheltering-concealing in sustaining," in which "originary emptiness opens, originary *clearing* occurs" (GA 65: 379–80/CP 265). This is our first hint that the "nothing" that pertains to be-ing is to be thought as the clearing that *opens* a way for and as timing-spacing-thinging.

Timing-spacing. In attempting to say be-ing and ab-ground I have already said much of timing-spacing. In short, timing-spacing, in the opening-extending that is the work of removal-unto and moving-unto, gathers things in their thinging, enowning them into their be-ing, clearing ways for them to arise and show themselves in their momentary uniqueness. In addition, consider this comment of Heidegger's, which places the thought of timing-spacing in the context of the first and other beginning: "Whence and why and how are both space and time together, for so long? What is the basic experience, the one that would not be mastered? (the t\here [*Da*]!) Only superficially, in accordance with the guiding beingness? But how [is] the 'and' [meant] for both? . . . the 'and' is in truth the ground of what is ownmost to both, the displacing into the encompassing open—an open that builds presencing and stability, but without becoming experienceable and groundable" (GA 65: 374/CP 261). Time and space belong inseparably together as timing-spacing, which is, as dynamic opening for the presencing of things, groundless and nonreifiable. Only in the emerging thinking of an other beginning does this—with some hesitation and difficulty—emerge into experiencing (*Erfahrung*), in which it becomes groundable (ab-ground as Ur-grounding). It is not only beings (things) that are displaced in the opening-extending of timing-spacing. Da-sein (i.e., us) is displaced into and as the t\here, the opening for the showing-forth (disclosing saying) of timing-spacing-thinging. This intimates, in a preliminary way, the emerging radical transformation of us in our relations to all things.

Thing. The thing things, that is, it gathers the fourfold of earth, sky, divinities, and mortals and is thus enowned in the displacing of moving-unto (the gathering that spatializes) and removal-unto (the gathering that temporalizes). Thus, thinging is timing-spacing, it is enowning, it is the ongoing dynamic relationality of be-ing. As be-ing is ab-ground, a thing is not a being. Any attempt to reify a thing has failed to heed the saying of the thing as thing in its emerging into the opening cleared for it in timing-spacing. As timing-spacing is also said as immeasurable nearing and intimacy, such failure to listen is a peculiar sort of deafness on our part, and all the more so as we are the t\here, the opening for this saying (GA 12: 199–203/WL 104–7).

Enowning. "Be-ing holds sway as enowning," or, to put it even more strongly, "be-ing is *en-owning*," which is "the temporo-spatial simultaneity for be-ing and beings" (GA 65: 13, 255–56, 260, 470/CP 10, 180–81, 183, 330). There is no be-ing apart from the enowning of things in timing-spacing. As enowning, be-ing is "unique and once only," arising as the uniqueness of each momentary, incalculable gathering (thing). This enowning, bringing each thing—each thinging—into its own, is the gathering that enables and is the saying (showing) of each thing in (and as) the opening that primal saying (enowning) clears. Belonging (*gehören*) with things in (and as) this opening, we are thus able to hear (*hören*) the saying of things and respond with our own saying-after them in language and in silence (GA 12: 242–49/WL 123–29).

Saying, as what is ownmost to language, is showing. But, as I just said, showing is what it is only in the enowning of things, which is the removal-unto and moving-unto of timing-spacing. And as enowning is another way to say be-ing, so too does saying "say be-ing" in its inseparability from and simultaneity with things. Saying, as enowning-showing, enacts our intimacy with be-ing, with timing-spacing-thinging, enabling us to heed the intimation of mystery, of what always withdraws, in things. Opening and attending to such mystery, we are able to keep it, to hold and ponder it in the thanc, the heart-mind that is open to the possibilities of thinking and dwelling.

Opening has been intimated already in several ways, as the t\here whereby Dasein is open for disclosing (and openness to mystery), as the opening-extending of timing-spacing, and, to say the same in yet another way, as clearing that displaces and shelters. Already in the first beginning, in the Greeks' wonder at the arising (*physis*) and un-concealing (*a-letheia*) of beings, opening is in play, though it was not thought explicitly as such. As Heidegger puts it, "By recalling [this] beginning, we come across the same thing, *the openness of the open*[,] . . . which is determined more essentially as *the clearing for self-sheltering*" (GA 65: 338/CP 236, Heidegger's emphasis; see also GA 65: 331–33/CP 232–33). But openness hints at emptiness. Just as that thought arises Heidegger cautions us against reifying openness as if it were akin to an empty container. Whether it is the opening-clearing that is thinging (timing-spacing) or the opening of the t\here, this is nothing extant; it is "not a being at all but is rather . . . the enquivering of enowning in the hinting of self-sheltering concealing" (GA 65: 339/CP 237). This is not to say that it is at all inappropriate to speak of opening as emptiness, though it is not "mere emptiness" (GA 65: 485/CP 341). In fact, returning to section 242 of *Contributions* gives us "emptiness" as another guideword for thinking.

Before we move into that discussion, look at what this gathering of the various guidewords that all say be-ing has yielded as reminder and as insight. And as saying is another way to say enowning, we are reminded that

the two modes of saying (in thinking—in the guidewords and their resonance—and in thinging) echo one another. I will say here again what I
said already in chapter 3, with some additional insight drawn from this
reflection on each of the guidewords, that what we are attempting to think
and bring to language is

1. gathering
2. . . . . of what is the same, belonging together
3. . . . . in its very dif-fering
4. . . . . in which each facet intertwines, echoes, and mirrors every other
5. . . . . in ongoing timing-spacing-changing
6. . . . . and thus refuses to be grasped in any definitive concept
7. . . . . or system of thought.

The last item calls for closer scrutiny. As I went through each of the guidewords above, saying them in terms of the others, it became quite clear that
they are all, indeed, ways to say various aspects or movements of the one
matter. It could seem, however, that since their relationships can be articulated with the relative degree of clarity that I have given them, perhaps
we are moving toward the possibility of constructing a systematic understanding of be-ing. Already in chapter 2 I pointed out that this would not
be the case. Why not? In the first place, these ways of saying be-ing do not
indicate, as I have said repeatedly, anything fixed, final, or subject to anything that would count as proof according to the history of Western philosophy. Systems belong to metaphysical thinking, thinking that attempts
to conceptually interpret our understanding of being and beings. In speaking of the joinings that structure *Contributions* Heidegger says that they
do not constitute a system. Neither do the joinings of the guidewords that
have so far emerged in our attempt to think and say be-ing. They open
ways and paths and provide directions for thought without establishing a
resting place—something akin to a theory, doctrine, or system (GA 65: 7,
81/CP 6, 56–57; ID 21–23/83–85). Be-ing "itself" can never—by this name
or any other—become an "absolute" around which such a system could
structure itself. "Be-ing is directly never definitively sayable—and thus never
only 'provisionally' sayable" (GA 65: 324/CP 460). That be-ing is not definitively graspable in language is something I have already pointed out. That
it—or any of the other guidewords—is not provisional either is a significant cautionary insight. What Heidegger is trying to get us to consider here
is another way in which our tendency toward dualistic, either-or thinking

can lead us astray in the attempt to think be-ing, to hold be-ing in question. It is not only the case that "be-ing" or "enowning" or "thinging" is not the last word or a definitive way to say the matter for thinking. They are not just provisional either, because that implies that there is or could be, at some point, a final, definitive word (or words) for the "simple one-fold." There will not be. Furthermore, that is not to be taken as a defect or lack in this way of thinking but, rather, its holding-to the matter, which is *ab-ground*. This is another way of reminding ourselves that the attuning of thinking in the other beginning is *reservedness*, which is "the creative sustaining in ab-ground" (GA 65: 36/CP 26). What is attuned in this thinking is not only the language that carries it but also the thinker. The only reason for carrying on with this thinking is its call, experienced as the joining of the thanc with the play of timing-spacing-thinging, which is, in our situation in relation to the first and other beginning of Western thinking, transformative through and through. The move toward system is a move to set the question of be-ing outside the heart-mind, at some objective distance, and thus evade the possibility of radical transformation. The thought of the *emptiness* of be-ing will not only add to our understanding but also serve to further hold thinking transformatively within ab-ground.

EMPTINESS

We can enter into the way Heidegger introduces the notion of emptiness into *Contributions* by thinking back to what was said of timing-spacing as moving-unto (the gathering that spatializes) and removal-unto (the gathering that temporalizes). Moving-unto says the "not" in be-ing in terms of the "not yet," while removal-unto says that "not" in terms of the "no longer." This, says Heidegger, is one way to begin to think the nothing that pertains to be-ing or, to put it another way, its nihilating in timing-spacing (GA 65: 288/CP 410). This nihilating is neither something positive nor something negative. In the first place, it is not somehow against or counter to be-ing but, rather, says something of how be-ing holds sway. "Only because be-ing holds sway in terms of the not [*nichthaft west*] does it have the not-being as its other. For this other is the other of itself. *Holding sway in terms of the not, it [be-ing] makes possible and enforces otherness at the same time*" (GA 65: 267/CP 188, Heidegger's emphasis). This otherness is—in terms of what becomes thinkable here—threefold. In the first place, we have the otherness within be-ing "itself," as it is always not-in-being, not-be-ing, always withdrawing in its very bringing everything into be-ing. In a sense, be-ing is always dynamically other *within* its own way-making, timing-spacing-thinging movement. The otherness of be-ing as timing-spacing-thinging also arises as the already-discussed nihilating of removal-unto and moving-unto, which makes possible the otherness that is the momentary uniqueness of each thing that arises. This reinforcing of the thought of the uniqueness of be-ing in each thing is significant. I emphasize the relationality of thinging and the "simple onefold" of timing-spacing-thinging. This is undeniable, and yet we must avoid the extreme that would want to take onefold relationality as mere "oneness." Things are not merged into some bland, uniform cosmic Jell-O. The situation is, in fact, quite the contrary. It is the thinking of "being" that attempts to capture what is general and common to all beings. It is the extremity of that thinking, where being functions as enframing, that enforces not otherness but functionally interchangeable uniformity (in standing reserve). The be-ing of things in timing-spacing-thinging, on the other hand, enables and maintains the unique otherness of things in their dynamic relationality with one another. They are the other of each other as well as the other of themselves (in that they never actually "are" but are continually changing).

I said that the otherness of be-ing is threefold. So far, however, I have only discussed two facets of the matter: the otherness of be-ing within itself (as not-being that always withdraws) and the otherness of things that arise and come to be-ing, differing from each other as well as from what they were and will be. The third sense is not separate from these but arises from the fact that be-ing "is" timing-spacing-thinging. There is no be-ing apart from thinging, from things. So the otherness of be-ing within itself is also the otherness of things as dynamic and in relation to one another. This we could call the otherness of be-ing as *other-ing*, which is a way to bring together these facets of our thinking of its dynamic: the play of gathering and dif-fering, of gathering as removal-unto and moving-unto, and of be-ing as not-being in manifold senses.

The nothing that pertains to be-ing is neither vacuous voidness nor a reifiable void. It is not simply a matter of something that is not occupied or engaged with something else but of timing-spacing-thinging, which is nihilating and other-ing in the senses just discussed. This, in turn, is another way to more fully think *opening*. Heidegger says that the emptiness of be-ing is *en-opening* (GA 65: 265/CP 379). The prefix strengthens the dynamic character of what we hear in the word "opening." The emptiness of be-ing, unlike a void, is always in motion, opening and clearing the way for (and as) timing-spacing-thinging. We know already that this emptiness, this energetic opening and clearing, is also thought as ab-ground. The staying away of ground is no mere vacating of some domain where there could have been a ground but, rather, "hesitating refusal" of ground (GA 65: 380/CP 265). That intimates once again the Ur-grounding that takes place as the threefold other-ing in the emptiness of be-ing.

There is yet another thought that prevents us from taking emptiness or nothingness to either extreme (voidness or reification). That is the fact that *this* emptiness is also always *fullness*. Well, of course it is. How could it be otherwise, since this emptiness is the nihilating that takes place as removal-unto and moving-unto, which is the play of the rich relational dynamic of thinging. As Heidegger puts it, "Fullness is the turning in en-owning," the other-ing that leads him to say in another place that emptiness is "actually the fullness of . . . what holds to ab-ground" in the counter-resonance of moving-unto and removal-unto, where we encounter the "not-character of be-ing as enowning" (GA 65: 267, 268, 382/CP 188–89, 266).

In short, the nothing that pertains to be-ing is the emptiness of be-ing thought as timing-spacing-thinging, which is, as the simultaneity of be-ing and things, (1) the nihilating of the no-more and the not-yet, (2) the other-ing of be-ing in itself (ab-ground) and of things (which are, of course, not two different "otherings"), (3) opening and clearing for and as timing-spacing-thinging, and (4) the fullness of the enowning of things. And so emptiness is another of the guidewords for the thinking of the first and other beginning, in joining with those that have already arisen in the course of thinking and those that are yet to come. Just as do the others, it has multiple meanings that shift and emerge in the various ways of playing forth the thinking along with and through them. One of the senses of emptiness is that of the ab-ground character of be-ing. But this, in turn, also resonates with our situation in the transformative thinking of the first and other beginning. In my first discussion of that context for thinking, in chapter 1, I pointed out that we get the first hint of ab-ground in the abandonment of being that comes to light in enframing. The first shock of that awareness begins to attune the thinking of the first and other begin-ning which, as it unfolds, is attuned by the reservedness that is in tune with be-ing's manifold withdrawing. This manifold withdrawing is not simply identical with the other-ing that enacts be-ing's emptiness but is in very close joining with it, saying somewhat different facets of the same com-plex movement (simple in its dynamic, very complex for thinking). As thinking carries on, drawn by the very questionableness and difficulty of bringing the matter to language or, perhaps better said, of following the movement that becomes possible in language, this attuning in reservedness becomes more and more compelling. Why do I say that? Heidegger indicates that this attuning is able to carry out its work because it arises within what is already "the most originary belongingness" (GA 65: 382/CP 267).

As early as chapter 3, in the discussion of timing-spacing-thinging, and then again with much more emphasis in chapter 4, I raised the question of our situation in relation to things (which is also to say, of course, in relation to timing-spacing, to be-ing, to enowning). The originary belonging to the emptiness of be-ing, which we experience as be-ing attuned to it in the reservedness that guides this thinking, is no mere abstraction for thought. It shapes everything that has been thought here and what may come. That statement about our originary belonging to the emptiness of be-ing occurs in the decisive section 242 of *Contributions*, "Time-Space as Ab-ground." The last few pages of that section bring its discussion of timing-spacing home within the thanc and make quite clear, once again, that the point is not a better understanding of time and space but transformation, of awakening new ways of experiencing (*Erfahrung*) things (GA 65: 388/CP 271). We can be attuned to be-ing's emptiness (fullness!) only because we already belong within it. At the very least, we see this in terms of our being of the fourfold that is gathered in thinging. Further, we can see this in that we have done the thinking to this point. Its difficulty (and if it were only a matter of academic philosophy, its uselessness for metaphysical, epistemological, or ethical theorizing) indicates that there is something here that draws us, something that moves and calls not just to our intellect but to our heart-mind, the thanc. Heidegger suggests that there may be here a "remembered awaiting (remembering a concealed belongingness to be-ing, awaiting a call of be-ing)," which echoes what he says of the way the thanc embodies and keeps what is to be thought (GA 65: 384–85/CP 268–69). How can we make sense of this call arising in our own heart-mind? Only insofar as we understand the nature of the belonging that allows it.

As I suggested in chapter 4, I can see no good (nonarbitrary) reason why, when we can see that be-ing and things cannot be reified, we should continue to reify ourselves. We quite obviously arise and belong squarely within the relational dynamic of timing-spacing-thinging. This does not simply equate us with things, though we are thingly. In fact, it is our arising in timing-spacing-thinging that pulls away from any such reductive move. Each moment of thinging is *unique*, since things are not simply presences or uniform beings. Likewise, so too are we unique in *at least* three senses. (1) We are unique individuals, even though, at present, enframing would like to efface this fact. (2) We are unique in the sense in which "momentary uniqueness" is meant in *Contributions*: we are dynamic, always changing, so that each nonlinear, noncalculable moment of enowning gathers and differs from all the others. (3) As humans, as Dasein, we are uniquely endowed—as far as we know—with the capacity to carry out this thinking. It is that third sense of our uniqueness, which draws on the other two, that must be considered in much greater depth if we are to fully understand the nature of dwelling with things in lucid awareness of their emptiness. We are not yet ready to leap forward into consideration of dwelling in that light. While I have brought in numerous paths of thinking by which to understand why be-ing and things cannot be reified, I have only somewhat indirectly supported my thought that we ourselves are also nothing to be reified, based on what can be found in Heidegger's work of thinking. There is much more that can and must be said about that. However, it calls on us to look farther afield for some suggestions on how to proceed. What are we asking here, precisely? It is the question that I brought in from Heidegger near the beginning of this chapter, which is actually two inseparable questions: to think *who* and what we are, as a question, requires asking *whether we are*. That is, it requires us to think seriously about whether we are beings, whether we are, in fact, anything that can in any way be refied. The two questions are, Heidegger said, inseparable, which means that the answer to either depends on the other. So it is not the case that we can work out an answer to the one and then the other. They are thought together.

It is no doubt obvious already that I intend to bring us around to the point where it is clear that we are not reifiable; what needs to unfold is the response to the "who and what" question that unfolds along with that. As those who inherit well over two thousand years of Western philosophy and religion, we carry around with us various presuppositions that would tend to lift us out of timing-spacing-thinging by retrieving some part or aspect of ourselves that can be reified or granted the standing of substantial self-existence. Clearly, the body—taken in a limited, strictly physiological sense—is not what most people think of here. We have a long history of thinking of ourselves dualistically, as being "body" and "mind" (or "soul"). When we think of this in terms of epistemology and of action in the world, we tend to play this dualism out in terms of subject (mind or, in some contexts, simply the human) as over against object (the things "out there," sometimes including our own bodies). The difficulty of letting go of the tendency to reify ourselves is not based on just some vague, general

fear of nonbeing but, rather, on our long-standing, unquestioning accept-
ance of dualism, along with the fact that this received presupposition fosters at the same time a belief in human superiority. I know from several
years of teaching environmental philosophy just how tightly intertwined
and tightly held these assumptions are. To question either idea, human
superiority or the substantial self-existence of the mind or soul, is often
taken as an intolerable attack on our own "special place" in the universe.
The fact is, of course, that in the thinking of the first and other beginning
the uniqueness of Dasein is emphasized (much more, in fact, than what
I have done in this book, though that will come in chapter 6). But right
along with that, all those old presuppositions are called into question.

Every last one of them.

What I have brought in so far are a few hints in that direction, hints that
emerge here and there in Heidegger. If they are brought together, their
hinting seems rather stronger, becoming more like a directive for thinking.
In the discussion of space that emerges from the account of thinging in
"Building Dwelling Thinking" Heidegger said that our mobility in time-
space tells us that we are not just encapsulated bodies. This suggests rather
strongly that we, like things, are relationally dynamic, that we, too, participate in timing-spacing-thinging, and not just as "mortals" in the fourfold.
The account of the thanc in *What Is Called Thinking?* tends to undermine
mind-body dualism in that it emphasizes the importance of being-moved,
so that what thinks is more appropriately thought as heart-mind and not
just as mind in the traditional Western sense. He also, in the same place,
says that what we think comes to us because of, or in, *contiguity* or contact with things (WHD 157/WCT 144). This resonates with and reinforces
and strengthens what we just read in *Contributions* about our being able
to think the emptiness of be-ing because it calls to us, due to our already
belonging to it.

Regarding subject-object dualism, a matter that must be taken up decisively, there is no doubt at all about where Heidegger's thinking leads. In a
conversation recorded between Heidegger and a Buddhist monk the discussion turns to the way that subject-object dualism hinders the unfolding
of thinking, but the dualistic presupposition is so powerful that it effectively
captures and imprisons our thinking. Heidegger says in that connection that
his "whole life's work . . . has been devoted to freeing us from this prison."[1]
That is rather stronger than a hint. As the discussion unfolds, the monk
draws from Heidegger the suggestion that the way to overcome this hin-
dering dichotomy is by way of releasement toward things and openness to

mystery, which means, in the monk's words, to gather oneself to the "nothing that is not nothing . . . [but] fullness." In response to that comment Heidegger says, "This is what I have been saying throughout my whole life."[2]

The direction is clear; the question is how to proceed. The inseparable play of emptiness and fullness, as indicated in Heidegger's conversation with the monk, has long been a core teaching of much of Buddhism. Furthermore, the Buddhist thinkers have not just recently begun to think in these terms. They have, during the more than two thousand years that we have been under the sway of metaphysical thinking, developed and refined and practiced ways of thinking and experiencing and dwelling with things— and ourselves—understood as emptiness. Perhaps they can help us out. This move will come as no surprise to readers with some background in Buddhist thinking, since the similarities between Heidegger's thought and Buddhist thought are so striking as to be inescapable.[3] However, the key here is for this to remain *thinking*, not just "comparative philosophy." We are attempting to think be-ing in the first and other beginning, opening up *its* possibility of radical transformation.

Radiant Emptiness

Before carrying on I should point out that what we can see of Heidegger's own sense of the relationship of his thinking to Buddhism is ambiguous. On the one hand, he was aware of the similarities and was open to dialogue, as indicated not only in the interview with the Buddhist monk noted above but in his own writing in "A Dialogue on Language," in *On the Way to Language*. In that dialogue it was particularly the notion of emptiness as understood by the Japanese that was of interest to him (GA 12: 97, 101–4/ WL 14–15, 18–20). On the other hand, there is a comment in *Contributions* that should make us pause, especially since it occurs in a brief discussion of the very topic I am attempting to think here: the insight arising in the thinking of the first and other beginning, that we are not "beings." Heidegger says,

> The more exclusively thinking turns to beings and seeks for itself a ground that exists *totally as a being* . . . the more decisively philosophy distances itself from the truth of be-ing. . . . But how is . . . renunciation of metaphysics possible without falling prey to the "nothing"? . . . The less a being man is and the less he insists upon the being which he finds himself to be, so much nearer does he come to being. (No Buddhism! The opposite.) (GA 65: 170/CP 120)

What are we to make of this? In 1938 Heidegger obviously did not find any deep affinity between the nothing that pertains to be-ing, the emptiness of timing-spacing-thinging, and what he thought the Buddhists meant by those words. By 1954, the date of his writing "A Dialogue on Language," his thinking had apparently shifted somewhat toward understanding that there was more resonance than he had thought. But he never, as far as I can tell, was as open to a deep encounter with the thinking of Buddhism as he was to that of Taoism. Why not? His conversation with the monk, mentioned above, contains a hint.[4] Essentially, Heidegger shared a very commonly held misconception about Buddhism. Heidegger and the monk discuss the Western view that takes our having language and thought as what makes us human. Heidegger indicates in the course of that discussion that "in contrast to Buddhist doctrine, Western thinking draws an essential distinction between humans and other living beings, such as plants and animals. Humans are distinguished by their knowing relation to being."[5] While Heidegger would be quite right in assuming that Buddhism makes no distinction in the *value* of different kinds of sentient beings (something with which it is unlikely he would disagree, given the way he problematizes the whole issue of "value," as we will see in chapter 6), he is incorrect in inferring that that means there is no significant distinction made at all between humans and the other sentient beings. In fact, the distinction made in Buddhism is precisely what Heidegger mentions: the capacity for a knowing relation to that deep mystery that goes by the names of being, be-ing, *dharmakaya,* buddha-nature, and so forth. In Vajrayana Buddhist teaching this distinction is thought as the eighteen endowments of a precious human life. I think it significant that this list is not an essentialist characterization of what it is to be human but, rather, a description of distinctively human possibility at its best, which is something quite often unfulfilled. Thus, the eighteen endowments include not only simply "being a human" (five of the eighteen pertain to that) but matters of inner and outer context, such as having full use of our sense faculties (two), having a disposition to be open to mystery and transformation (three), being fortunate enough to have the time and appropriate situation to think about spiritual or philosophical matters (one), and living in a place and time where there is help and guidance for our thinking and practice (seven).[6]

If Heidegger had been aware of this basic Buddhist teaching, it is much less likely that he would have said, "No Buddhism!" That said, I have no intention, in any of what follows, to simply equate Heidegger's thinking with Buddhism. I introduce some fundamental teachings of Buddhism in the spirit of bringing another significant *joining* into play with the thinking of the first and other beginning, particularly insofar as it can help us to think more clearly and carefully about the emptiness of be-ing as well as the ramifications of this emptiness. The guideword for the discussion is this passage from the Prajñāpāramitā Hridaya Sutra, popularly known as the Heart Sutra (its longer title literally means "heart of perfect wisdom").

Form is emptiness.
Emptiness is form.
Form is none other than emptiness.
Emptiness is none other than form.
In this same way feeling, perception,
Mental formation, and consciousness are empty.
Thus, Shariputra, are all dharmas [things] emptiness.[7]

Notice that it is not just other things that are said to be empty but also the things that go to make up our own human sense of existence: mental formation and consciousness in addition to form, feeling, and perception. This echoes the insight that goes all the way back to the historical origins of Buddhism. Hinduism, from which Buddhism diverged, was grounded on the notion of Brahman, which is pure, absolute, eternal, and unchanging *being*. Hindu practice aimed at removing the mundane obscurations that cover and hide Brahman, revealing the Brahman that is there all along, only hidden by the obscurations. Brahman thought in its hiddenness is called the atman.[8] One of the earliest Buddhist insights was named in this word: *anatman*, that is, "no atman." But since atman is essentially another name for Brahman, the word also means "no Brahman." So from the start Buddhism moves away from the notion of "being," whether as ground of beings—in which case, we can, without misinterpreting the matter, think of *anatman* as ab-ground—or as characterizing either our nature or that of the other things in the world. As Buddhist philosophy developed to support Buddhist practice, *anatman* began to be thought more broadly as emptiness, that is, not so much in terms of negating a specifically Hindu idea but as a central guideword for Buddhist thinking itself.

Fairly early on, in the Abhidharma teachings (which can to some extent be thought of as encompassing basic Buddhist epistemology and psychology), the thought of *anatman* was brought into relation with the teaching of the five *skandha*s, or aggregates, that come together in each of us. These are the very things (form, feeling, etc.) listed in the Heart Sutra. Quite awhile before the Prajñāpāramitā literature existed, though, the Abhidharma thinkers used the five aggregates to reinforce the insight that we are not beings, that we are not substantially self-existent entities. The reasoning is that, since our very existence depends upon these components, each of which is itself subject to changing and is dependent on many causal conditions for its existence and functioning, there is no justification for us to consider ourselves as substantially existing entities. In short, we are not beings in either the Hindu sense or the Western metaphysical sense. Does that mean we simply *do not exist* at all? Early Buddhism offered various responses to that question, at times taking the matter to that extreme while at others asserting that perhaps the aggregates themselves actually existed.[9]

It was the emergence of the Prajñāpāramitā Sutras, along with the work of the great Madhyamaka (Middle Way) philosopher Nāgārjuna, that set the stage for a radically and decisively different way of thinking about such matters. The heart of his philosophy is the exposition of the absurdity of what are called "the four extremes": being, nonbeing, both being and nonbeing, and neither being nor nonbeing. If we look carefully at those four, it is apparent that they comprehensively encompass all the possibilities of metaphysical conceptualization. The core of the Heart Sutra captures the heart of Nāgārjuna's insights. I will take a closer look at them, point by point.

Form is emptiness. Anything that has form depends on many causes and conditions for its existence. We can think of these as the five aggregates of Buddhism or in terms of the fourfold of timing-spacing-thinging in Heidegger. In either case the result is the insight that everything that arises into appearance is empty of substantial, independent self-existence. Notice, too, that the Sutra says that form *is* emptiness. It is not just akin to emptiness; it simply is empty. There is no dualism here of two things or beings or qualities that could be named as form and emptiness. Another crucially important insight is that emptiness here cannot in any way be understood nihilistically. Just as is the case with Heidegger's comments on the "nothing that pertains to be-ing" and the emptiness of timing-spacing-thinging, so it is here. The situation is somewhat akin to the manifestation of a rainbow. It is "really there" in the sense that we and others see it, but it is entirely dependent on light, moisture, and the precise angle of the perceivers. And no matter how determined we might be to grasp it and fix it in place so that we can enjoy its beauty permanently or even just a little while longer, that is not possible. The Buddhist insight is that this is true of everything that depends on causes and conditions, everything that arises in the play of dynamic relationality. Thus it is that form (and all the other aggregates as well as the "beings" they gather into) is emptiness. The very idea of "being" is also, according to Nāgārjuna and later thinkers, itself empty. It is only a concept. That is precisely what I said of "being" in chapter 1 in the context of the thinking of the first and other beginning of Western philosophy.

Emptiness is form. Think again of the rainbow and all that comes into play every time we see one. Think of the loaf of bread I described in chapter 2 or the tomato sandwich in chapter 4. Emptiness is quite obviously not annihilation. What Heidegger says in section 242 of *Contributions* Buddhist thinkers from Nāgārjuna forward have also said. "Emptiness . . . is not just a blank, dark state . . . [but] the fullness of all qualities."[10] Emptiness is not to be taken as some kind of absolute at which to aim in either thinking or practice. Emptiness and form are mutually dependent ideas that help us think about the way things gather and manifest and show themselves. The only way that Buddhist thinkers have differed from Heidegger on this point is in being more clear right from the start that this applies not only to what we usually call things but to us as well.

Form is none other than emptiness. Neither form nor emptiness somehow exists in itself. So in saying that form is emptiness or that emptiness is form we are not thinking of a relationship between two entities. Mahayana Buddhism thinks timing-spacing-thinging in terms of the dynamic play of interdependent arising or co-originating (Sanskrit *pratītya-samutpāda,* Tibetan *tendrel*). Things are empty. So is "emptiness" empty. It is just a way of thinking more clearly about things. Without things there would not even be an idea of something that we could call emptiness. As Heidegger said, be-ing as ab-ground does not somehow exist beyond beings (things). And, as Nāgārjuna taught, there is no sense in positing the existence of "both being and nonbeing."

Emptiness is none other than form. This appears to repeat what is said in the previous line of the Sutra, but in the light of the relationship between Nāgārjuna and the Prajñāpāramitā Sutras there is a subtle but important thought that needs to be added. "We have reached the level of understanding that the ideas themselves of existence and non-existence are faulty and extreme . . . [but] we have still not transcended clinging, for the idea that existence and nonexistence do not exist is still an extreme position. . . . The argument subverts our hold onto any position whatsoever. . . . The true nature is beyond words, beyond the limits of our imagination. Otherwise we could define it."[11] So it is that the proposition "neither being nor nonbeing exists" is as untenable as "being," "nonbeing," and "both being and nonbeing." The upshot is that any possible way of conceiving of things *metaphysically*, in terms of beings, is untenable. And since, as Heidegger also points out, concepts themselves are ways of grasping and reifying, that means that the way things arise and show themselves is, at bottom, inconceivable. Phenomena themselves are *ineffable*, inconceivable in terms of any metaphysical way of thinking them and bringing them to language.[12] And yet there they are, appearing all around us. As the great fourteenth-century yogin and philosopher Longchenpa puts it, all things are "forms of emptiness, clearly apparent yet ineffable."[13] We already know that Heidegger thought timing-spacing-thinging as empty, according to what he says in section 242 of *Contributions*. If we take up once again his simpler, more poetic account of the same matter in "Building Dwelling Thinking," we can think it in terms of emptiness as well. Earth, sky, mortals, and divinities are empty of substantial self-existence, whether we think of them as naming the large kinds of things Heidegger describes or whether we think of individual facets such as stones (earth), clouds (sky), myself (mortals), or a particular hint of mystery glimpsed when the doe's tail flashes white as she leaps over the fence (divinities). As gathering this fourfold, thinging is the in-itself-empty gathering of "forms of emptiness," to borrow Longchenpa's phrase.

What, then, is the point of all this philosophical thinking about emptiness? For both Buddhism and Heidegger the point is to open us to the possibility of transformation. Buddhism, from its beginning, saw the origin of much suffering in our dualistic, reifying thinking that grasps and clings to beings (including ourselves) in spite of their ineffable, ungraspable, ever-changing nature. Heidegger is deeply concerned with the likelihood of our being trapped in the most extreme form of metaphysical thinking, enframing, with all its ramifications for the constrictive narrowing of our thinking and ways of dwelling, as I discussed in chapters 1 and 2. Whether you come at this matter from the direction of the thinking of the first and other beginning or from the direction of Buddhist thinking, reification (along with all the dualistic divisions and barriers that spring from it) is at the heart of much that is not only questionable but also conducive to a great deal of misery and needless suffering. And it is, in contrast, the insight into the emptiness of timing-spacing-thinging that opens ways to think and move outside the limits of reification.

Evoking the thinking of Nāgārjuna, contemporary Buddhist teacher Thich Nhat Hanh says, "Thanks to emptiness, everything is possible. . . .

If I am not empty, I cannot be here. If you are not empty, you cannot be there. Because you are there, I can be here. That is the true meaning of emptiness. . . . Emptiness is impermanence, it is change. . . . [W]ithout impermanence nothing is possible."[14] As I have already suggested, the realization that timing-spacing-thinging is empty, through and through, does not result in the disappearance of things. In Buddhist terms nirvana, which is freedom from samsara (habitually clinging to reifying, dualistic ideas and perceptions), does not mean that samsara disappears or that we actually move from one metaphysical territory to another. Language akin to that of the Heart Sutra is appropriate here. Both samsara and nirvana are empty, so samsara is none other than nirvana, and nirvana is none other than samsara. It is our thinking and perceptions that have changed; letting go of clinging to reifications transforms samsara into nirvana through transforming all our relationships to and within it. Another way to say this is that since samsara was only our idea, so is nirvana. But the empty-full arising of things continues, just as it always has, in all its dynamic energy. As Longchenpa put it, "The division between samsara and nirvana collapses—not even basic space exists innately. There is no . . . 'How is it?' 'What is it?' 'It is this!' What can anyone do about what was so before but now is not? Ha! Ha! I burst out laughing before such a great marvel as this!"[15] And another teacher says, "If we were to condense *The Heart Sutra* down to an even more succinct message, it would be the single syllable Ah."[16] Heidegger, too, emphasizes that in the thinking of the first and other beginning we are not simply shifted from one domain to another (say, from "metaphysics" to "nonmetaphysics"), nor is it the case that "being" actually *existed* apart from our reifying of it so that it could now fall out of existence. Instead, what we experience is an ongoing resonating-back-and-forth in the thinking of the first and other beginning. Each resonant movement, however, does not simply return to where it was but transforms the thinking, and as it does it also changes us, the thinkers.

There is another way to realize, from within the thinking of the first and other beginning, that the emptiness of be-ing, of timing-spacing-thinging, does not mean the disappearance of things. On the contrary, it is the Ur-ground of thinging. To put that another way: if things (thinging) were not empty, there could be no things. Emptiness—as one of our guidewords—names another facet of the Indra's net of thinking. In joining it deepens our understanding of several of the other guidewords. "Emptiness" says the opening-extending of the play of removal-unto and moving-unto in and as timing-spacing-thinging. If we could imagine a fixed, unchanging, nonrelational being somehow entering into the dynamic opening and clearing, the result would be like the Ice Nine in Kurt Vonnegut's novel *Cat's Cradle*. Ice Nine was an unnatural substance that instantly and irrevocably froze everything that came into contact with it. Those things then effectively became Ice Nine, freezing everything they touched, and on and on until there would, eventually, be nothing but Ice Nine: ultimate stasis.[17] Philosophically speaking, this would be the ultimate perfection of the concept of being: purely constant presence. "Being," thought carefully, is rigid closure, cutting off the possibility of change and of relationship. Strictly thought, if "being" actually existed, there could be no beings, not if by "beings" we mean the phenomena we experience as well as ourselves. There would only be "being" in its pristine, unchanging, nonrelational oneness, just like the final result of Ice Nine.

Be-ing, on the other hand, says *openness* of and within ever-changing dynamic relationality. Because the word "be-ing," however, is ambiguous and in constant tension with "being," neither Heidegger nor we simply rest on that word but instead call into play the entire ever-shifting, shimmering, and growing net of guidewords that remain in resonant intertwining transformation. We do not grasp at some unattainable experience (*Erlebnis*) of "being" but remain open to experiencing (*Erfahrung*) the dynamic of be-ing, which is, after all, where we already are, though we are only now beginning to find it thinkable. This fresh in-grasping of our emerging within timing-spacing-thinging enables us to move with the turnings in enowning and to enact transformative thinking. It is the incipient awareness that, as Longchenpa puts it, "*Experience is open-dimensional.*"[18] We, in our own ineffable bodies-minds and in our heart-mind, the thanc, are inseparable from the open play of timing-spacing-thinging. This is, I think, another insight into the difference between grasping and in-grasping, a distinction used by Heidegger that I first introduced in chapter 2.

There is another array of guidewords that can help us better understand how our inseparability from the empty and open dynamic of timing-spacing-thinging enables the possibility of our in-grasping and thinking it. Longchenpa, in discussing the "simultaneity of awareness and emptiness," goes on to say that this naturally occurring awareness is "empty yet *lucid* . . . without basis . . . with all that manifests being clearly apparent yet ineffable."[19] Longchenpa speaks of this awareness as lucid and brings the point home in saying that "uninterrupted openness is naturally radiant and naturally lucid, unconstrained by reification."[20] The radiant emptiness of timing-spacing-thinging resonates and echoes in us, enabling our awareness of it. Can we take that awareness further and *think* it without falling prey to reification? This once again calls on us to be aware of our relationship to and use of (or, as Heidegger might sometimes put it, our being used by) language. To be "unconstrained by reification" calls on us to continually remember to practice releasement toward things, as I described it in chapter 2. What we encounter now is the convergence of releasement toward things and openness to mystery. Releasement of those learned, habitual tendencies to grasp and reify our ideas does not leave a gaping void in its place. Thinking, in and as the thanc, enters into "where it already is," the way-making dynamic of opening-extending timing-spacing-thinging. We have read where Heidegger also refers to this opening as *clearing*. This occurs in several places in Heidegger, usually by way of the word *Lichtung*. Even to native English speakers the fact that there is a sense of "lighting" carried in this word is evident. This resonates with the lucidity and clarity that Longchenpa speaks of. The clearing that says something of opening is "lighting" in a double sense: the obvious lighting up that says arising into the possibility of appearing and also "lightening." Heidegger helps us to understand this crucial point in "The End of Philosophy and the Task of Thinking."

> The adjective *licht* "open" is the same as "light." To open something means: To make something light, free, and open. . . . The openness thus originating is the clearing [akin to a clearing in the forest]. . . . Light can stream into the clearing, into its openness, and let brightness play with darkness in it. But light never first creates openness. Rather, light presupposes openness. However, the clearing, the opening, is not only free for brightness and darkness, but also for resonance and echo, for sounding and diminishing of sound. The clearing is open for everything that is present and absent.
>
> It is necessary for thinking to become explicitly aware of the matter here called opening. We are not extracting mere notions from mere words. . . . Rather, we must observe the unique matter named with the name "opening," . . . free openness, [which] is a "primal phenomenon." . . . The phenomenon itself, in the present case the opening, sets us the task of learning from it while questioning it, that is, of letting it say something to us. (TB 65–66)

Opening: clearing: lighting: lightening: freeing. And if we think back to what I did earlier in this chapter, saying "opening" in terms of many of the other guidewords that were in use up to that point, we can also see how this adds something very significant. There I said that opening is

(1) the opening-extending of timing-spacing-thinging, (2) the clearing that displaces and shelters, (3) emptiness, and (4) *us*, though I did not use that word. Again and again we need to remind ourselves that what is going on here is multifaceted transformation. It converges on us and our relationships with thinking, language, things, and the mystery of timing-spacing-thinging. What I actually said earlier in this chapter (the "us" of point four above) is that one way Heidegger says the open is by calling it the t\here, the *Da* of Dasein. All the way from *Being and Time*—where the t\here names our being open for disclosure, which through and through characterizes Dasein—on, this comes up in Heidegger's work at key moments.

   "Da-sein never lets itself be demonstrated and described as something extant. . . . The t\here [*Da*] is the open between that lights up and shelters—between earth and world, the midpoint of their strife and thus the site for the most intimate belongingness" (GA 65: 321–22/CP 226). What does this say? The first thing we know quite thoroughly by now: Dasein is not a being. Already this shows how the thinking here has moved forward from the first attempt to place the question of the meaning of being in *Being and Time*. We are t\here—neither just "here" nor "there" in terms of parametric space—*as opening*, as a midpoint in timing-spacing-thinging (the strife of earth and world). We belong t\here, intimately. This opening that we are, however, is not just a blank, passive "letting something happen." Just as is the case with opening thought as points one, two, and three in the previous paragraph, our opening is always relationally *dynamic*. Heidegger says here that the t\here is the "open between that lights up and shelters." We light up and shelter things, we light up and shelter timing-spacing-thinging, we light up and shelter the mystery of their ab-ground open-dimensionality as and *when we think*. We light up and shelter things, thinging, and their mystery *when we dwell*, when we stay with things as things, caring, preserving, and sheltering and *freeing* them. Only if we can free ourselves from the confining nets of reifying dualism through in-grasping the emptiness of timing-spacing-thinging can we begin to understand what it might mean to free things as well.

   Once again we encounter the thought of the possibility of the thorough-going transformation of our relationship to, well, to *everything*. So we carry forward with the attempt to think this, of undertaking the task that opening itself sets for us, in *letting it say something to us*. In chapter 2 we thought along with Heidegger as he drew out the thought of the heart's core of language, saying as showing. This saying is not just something pertaining to language, however, but arises in timing-spacing-thinging itself as things "say themselves" to us. This silent saying of things, bringing together language and enowning, language and the open, is, says Heidegger, "measure-setting in the most intimate and widest sense" in that it is our grounding as Dasein in the midst of timing-spacing-thinging (GA 65: 510/CP 359).[21] The radiant clarity of empty timing-spacing-thinging says itself to us in what is readily brought to language, in what is intimated and only with difficulty brought to language, and in what must be encountered in silence that is nonetheless the most telling of all. How are we to dwell with things in openness to this mystery, in openness in which we free ourselves as we free things? That is the question I carry forward into chapter 6.

6

Staying with Opening

How are we to dwell with things in openness to the mystery of their (and our) timing-spacing-thinging, enacting our mutual freeing? Any attempt to even imagine an openness that is genuinely *freeing* is at this moment hedged about with constraints that at times seem unavoidable. In chapter 4 I mentioned the ambiguous energy of language, which can echo the saying and showing of things or can entangle us so tightly within enframing that there seems to be no way out. We are continuously bombarded with the productions of contemporary information technology. We find ourselves strongly influenced and shaped by the things presented to us by way of news reports, popular science with all those studies telling us how to improve our chances of avoiding an early and gruesome death, television and film shows, advertising, and the internet. This is simply a fact of much of our day-to-day experience. In terms of the thinking of the first and other beginning it is not the least bit difficult to see in this one of the primary means by which enframing channels us into ever-increasing uniformity.

It is not just a matter of *what* is being presented to us. If anything, the *way* this information comes to us is even more powerful in its ability to mold and constrain us. The increasing computerization of language, based on reducing it to the most simple of components (the on or off of an electrical impulse), shapes us in ways we are only just beginning to realize. Just as the Greeks' shift from oral culture to the then-new technology of alphabetic writing allowed them to open up an entirely new way to think and relate to beings, our current shifting into a culture of electronic information processing cannot but change us.[1] One of the changes that is most apparent is a generalized "speeding up" that continually presses in upon us. We often simply do not seem to have enough time to do what we think we must do. I notice this frustration in my students. We hear about some of the results of the anger that arises in a speeded-up world in accounts of road-rage incidents. We see this compulsion toward speed in the popularity of fast food, prepackaged processed foods, and instant anything. In writing this book, attempting to *think* about all of this in such a way that others can also be provoked to think deeply about it, I became aware of this speeding-up as it manifested in intense anxiety about "meeting the deadline." Thinking is one thing. Using information technology to bring it to the printed page is quite another. The cognitive dissonance between what I am attempting to think and the requirements of book production was striking. Gazing at a computer screen for seemingly endless hours, I began to feel cut off from the earth, my family, and even my own body. If the mother woodchuck had begun her tentative approach toward me now, would I even have noticed? I doubt it. This is no mere complaint about feeling rushed but, rather, serves as a reminder to persist in thinking about the context and roots of our day-to-day experiences.

The early shift from oral culture to alphabetic literacy accompanied and fostered metaphysical thinking, the thinking of beings grounded on the constant presence of being. Our current shift from traditional literacy to high-tech information processing fosters and accompanies the culmination of metaphysics in enframing, the enactment of beings' loss of standing as beings or even objects. We, too, are in danger of losing ourselves in enframing in our very attempt to maintain control of it. We use language, and we use information technology, to try to maintain this control, even while the language of enframing makes use of us. Even the most impoverished of language says and shows something, if only by way of hinting at what is being avoided, suppressed, or covered over. In our current situation I would suggest that it is not just the information and language itself that must be questioned concerning what it says and shows but also the *way* in which it reaches and penetrates us. Contemporary information technology very nearly *forces us to be open* to its output. This openness, however, is not opening as clearing and lighting the way for the self-showing of things. The openness enforced by information technology is the openness of the passive consumer who becomes more and more akin to the technology itself. The technological information output is our input. We process this input, giving limited feedback and producing additional information—or informative actions, such as spending money—as our output.

Am I speaking too strongly when I say that this constant and fast-paced influx of technologically processed information *compels* us to be open receptacles for its input? Stop and think. Try to imagine a scenario in which, *for one week*, you are not available as a receptacle for what comes out of contemporary information technology. How easy is that to imagine? What would it take to carry out such an intention? No television, radio, internet, e-mail, compact discs. No newspapers, magazines, billboards. And don't forget food packaging; you would also have to refrain from reading the cereal box telling you that studies have shown that these crunchy bits are good for your heart. You would need to unplug the telephone, too, to avoid the telemarketers with their computerized dialing machines as well as call the post office to tell them to hold your mail, with its load of computer-directed mass mailings. I could go on, but this is probably enough to make the point. We are very nearly *forced* to *stay open* to the technologically produced information bombardment.

This compelled openness itself tends to cut us off from the opening that is needed for thinking and dwelling. It occupies very nearly our every waking moment. There is, it seems, *nothing left* in us by which to reawaken our own lucid awareness of our deep bond with the things around us or of the deep capacity we have to think and co-respond and care for those things. At an extreme the result is what Heidegger calls flight from thinking and thoughtlessness, which is having our capacity to think reduced to mere calculation and problem solving. At times we have a sense that there is more, but then our lives themselves seem to become another kind of problem to be solved. So, to counter our indoctrination in closing off, we need to (re)learn another way of opening. There is a struggle between the forced openness that tends to close us off and opening for thinking and dwelling. This opening is not the coerced passive openness of the information consumer. As we found early on, to open and to stay open for thinking and dwelling at first involves some unlearning.

To even begin we had to let go of our usual aim at acquiring better concepts, theories, and systems of thinking. Heidegger himself, starting with the question of the meaning of being, opened up a radically transformative *way* of thinking. When focusing with particular care on following through on the question of the meaning of being, he called this the thinking of the first and other beginning. That thinking both arises as and opens up what, most of the time, Heidegger simply calls *thinking*. I have attempted to think *along with* Heidegger, to see where that leaves us and where we might proceed, thinking *after* Heidegger. Reading and thinking with Heidegger has required entering into a mindful dialogue with him, interpreting what he says in the spirit of what he says of dialogue in "A Dialogue on Language." "Dialogue is determined by *that which* speaks to those who seemingly are the only speakers," human beings, whereby "the one thing that matters is whether the dialogue, *be it written or spoken or neither,* remains constantly coming" (GA 12: 143–44/WL 52, second emphasis mine). Thinking with Heidegger thus means engaging *the matter* for thinking as it is opened up by Heidegger. But as he quite clearly suggests, such thinking does not stay only there but must keep constantly coming. There is neither *arche* (as we are moving within ab-ground) nor *telos,* some final goal or end for the thinking. So thinking with Heidegger also requires both staying with his thinking and thinking after him, engaging in dialogue not just with him but with the matter, namely (to use one of its names), timing-spacing-thinging.[2] That is why he says that the dialogue may be written or spoken *or neither.* Listening and responding to the saying of things and to the withdrawing of be-ing in thinging may be carried on in silence. It might not even be brought to written or spoken language. It may be carried forward in gesture. It may involve action that stays with things to nurture, preserve, and free them, in which case thinking dialogue with the matter may be called *dwelling.*

Dwelling, as habitation, involves habituation. The releasement toward things that I have been emphasizing again and again is a matter of letting go of old habits that hinder engagement with the matter and thus tend to block thinking and dwelling. This involves not only those things that tend to block thinking at the start, mentioned just above, but also the ideas about the nature of things that we learn—more often by implication than directly—from when we are very young and first learning language. Thus far, in thinking very carefully about the relational dynamic of timing-spacing-thinging, it is the very notion of "being" that gets called into question and then released into the emptiness of ab-ground, be-ing. Reification is brought squarely to task as a hindrance to the thinking of timing-spacing-thinging. But how can this come into play day-to-day, in opening, in dwelling, in staying with things? This requires taking a much closer look at the transformative ramifications of the emptiness of be-ing.

Transforming Dasein

Throughout *Contributions* there are several different ways that the word "Dasein" is written: Dasein, Da-sein, Da-*sein, Da*-sein, and *Da-sein.* I want to explore some of what is going on with those shifts of emphasis as one way of entering into what I said in chapter 5: we are not beings, but, in one significant respect, just as other things, we are radiantly empty timing-spacing-thinging. There is that and there is along with and as that our unique enowning, which Heidegger also emphasizes. As I move forward with this exploration, remember what Heidegger said more than once as he began to engage a particular question: follow the movement of thinking without fastening tightly (grasping) on to particular concepts or relations between concepts.

In *Being and Time* Dasein, each of us as "being there," is said to be open as the t/here for the disclosedness of beings. That is one way of saying that Dasein is being-in-the-world. Although Dasein is "in each case mine," the t/here of Dasein is not some entity, even in *Being and Time*; it is temporalizing through and through. Dasein is its ways of taking up (or not) its relations to the beings of its world. This is, to say it another way, its temporalizing: finding itself attuned, falling into the midst of beings, projecting open its possibilities, and taking up discursive disclosure in silence or in linguistic articulation. I discussed this in chapter 1. Most of the time Dasein is said to be "inauthentic," with its ways of enacting its selfhood merely playing out the strictures and constraints of the they-self, accepting common notions without much question, wallowing thoughtlessly in prepackaged experiences (*Erlebnisse*), and thinking rather well of itself as a human over against all other kinds of beings. To be "authentic" means to mindfully own one's situation and possibilities with resolute openness and decisiveness. In *Being and Time* Heidegger is clear that the distinction between authentic and inauthentic Dasein is not to be understood in a moral sense or as implying greater or lesser degrees of "being." Dasein never simply shifts itself from inauthenticity to authenticity and remains there forever after (GA 2: 56–57, 167–70, 224–26/BT 67–68, 163–65, 213–14). In the hermeneutic approach of that text the distinction functions to lay out a broad range of Dasein's possibilities in its intrinsic dynamic relationality (temporalizing). Still, even for authentic Dasein the question of being is only just approaching over the horizon as a hint of what may become thinkable.

As the possibility of attempting to form and think the question of being comes to light we can see the thinking of Dasein moving into the crossing in(to) the first and other beginning. "Da-sein is experienced [*er-fahren*] . . . as Da-*sein*, enacted and sustained by a displacing shifting-into. This requires: sustaining the distress of abandonment of being" (GA 65: 309/CP

217). As the history of being culminates in its extremity as enframing and all beings begin to lose their standing as such, not only do we not know the meaning of being, we are hindered in many ways from inquiring and thinking about it. And even in the most serious and clear-headed attempt to think what "being" says, we find that "such an examination has nothing to hold" onto (WHD 137/WCT 225). The situation is ambiguous. Even enframing at its most rigid enacts a way of revealing beings, hinting at the mystery of all arising and revealing; in its very concealing in enframing be-ing's withdrawal begins to call out and draw thinking along after it (GA 7: 34/QT 33). "If the distress of abandonment of being is sustained, this sets into play the possibility of the deeper thinking and questioning, the grounding question of an other beginning: how does be-ing hold sway?" (GA 65: 78/CP 54).

How be-ing holds sway is in-grasped and said in many ways in that Indra's net of resonating guidewords in joining play with one another: enowning, timing-spacing-thinging, and radiant emptiness. And what is the situation of Dasein as enfolded in this unfolding thinking? I will repeat something said just above, adding what comes in the following paragraph in the text. "Da-sein is experienced . . . as Da-*sein*, enacted and sustained by a displacing shifting-into. This requires: sustaining the distress of abandonment of being. . . . *Da*sein's projecting-open is possible only as shifting into Da-*sein*" (GA 65: 309/CP 217). In the transformative thinking of the first and other beginning (1) Dasein is first enabled to place and think the question of the meaning of being, (2) thus being shifted into Da-*sein*, who sustains the distress of abandonment of being and keeps thinking, moving the question of the arising and holding-sway of be-ing, and only thus (3) projecting open *Da*-sein. It is clear that the third way of writing Dasein puts special emphasis on the t/here. The t/here: opening, disclosedness, temporalizing being-in-the-world. In the thinking of the first and other beginning, however, the thought of the meaning of the t/here is deepened. Already in *Being and Time* "the t/here does not mean a here and yonder that is somehow each time determinable," and now we can say also that it means "the *clearing* of be-ing itself, whose openness first of all opens up the space for every possible here and yonder" (GA 65: 298/CP 210). The t/here is the clearing and opening of timing-spacing-thinging. *Da*-sein is *Da*-seyn. (*Seyn* is how Heidegger writes be-ing, and so I suggest we might on occasion find it appropriate to write the word this way.) And so it seems, as we follow the movement of this thread in thinking, that we can detect a pattern in how the transformation of Dasein can be thought and said.

The pattern, expanded a bit based on chapters 4 and 5, could be laid out like this:

1. Dasein: temporalizing the disclosedness of beings but barely able to begin to ask the question of the meaning of being.
2. Da-*sein*: staying with the distress of abandonment of being, holding the question, and being shifted into crossing in(to) the first and other beginning.
3. *Da*-sein: questioning mindfulness of oneself as clearing and lighting, as opening, as belonging within timing-spacing-thinging, and beginning to listen for the saying of things *and* the withdrawing intimation of be-ing.
4. Da-*seyn*: persisting in the in-grasping of be-ing, thinking for the sake of dwelling.

Is this helpful? Yes it is, to some extent. We can find all of these thoughts in *Contributions*, and to list them in this way lets us see a bit more clearly some of the movement of the thinking of the first and other beginning insofar as it touches on the transformation of Dasein. However (and this should come as no surprise at this point), such a list flattens and systematizes the movement of thinking if we put too much emphasis on it. We have been so decisively warned off of systematizing this thinking by now that the flattening is probably the more serious danger.

We can, for clarification, think of the transforming of Dasein in this way, but then we must let go of grasping at it as a whole or in its elements. There is a pattern here, but it does not attach itself to each word on the list in quite that way. Heidegger himself is not particularly consistent in how he writes Dasein in *Contributions*. Here we need to call on that reminder about following the movement of thinking rather than attaching to the words that carry it. The path of thinking in the first and other beginning is not subject to being laid out as a linear, step-by-step progress toward some goal but is determined by being able to stay open to attuning to and by be-ing (timing-spacing-thinging). The problem of reification that I introduced in chapters 3 through 5 is quite often a matter of reifying our words, taking them as referring to beings. Or, as in this case, we could reify words by taking them as referring to characteristics of a being: Dasein. However, "Da-sein never lets itself be demonstrated and described as something extant. It is to be obtained only hermeneutically[,] . . . according to *Being and Time,* in the thrown projecting-open" (GA 65: 321/CP 226). So the question is not What *is* Dasein? but, rather, What does this word Dasein *say?*

Part of interpreting the saying of Dasein will be to take careful note also of what it does not mean. We have already two touchstones for interpreting the meaning of Dasein: (1) the t/here says *opening,* and (2) Dasein, which is not a being, should not, in the whole or in any facet, be reified. The second touchstone will help us rule out several things that Dasein does not mean, so I will take a closer look at what unfolds from that first and then return to a deeper inquiry into the matter of Dasein as opening.

Dasein is, says Heidegger, in each case *mine;* it says something about my self. But he also says quite bluntly that this self that emerges in the open that is the t\here "is never 'I'" (GA 65: 322/CP 226). How are we to make sense of this? This "I," written in just that way, invokes the ego of Western philosophy. It is the "I" of dualistic thinking. As long as we continue to think dualistically we will continue to reify ourselves, whether we are attempting to think in terms of Dasein or not. Letting go of reification requires letting go of all dualistic modes of thought. Dualistic thinking has, however, a very tight grip on us. It is not just something that frames philosophical discussions, which take place within dualistic paradigms of "intellectual life." It also thoroughly shapes popular culture and our day-to-day lives within it. So to let go of dualism calls first for reminding ourselves just how pervasive it is and getting some clarity about how it plays out in our thoughts and lives. I will examine these significant forms of dualism: subject-object, mind-body, and self-other. They are all interrelated, though for clarity we should look at them individually at first. They all arise from reification, and once they are in play they turn back to reinforce reification. All of them tend to dominate us, and all of them hinder the possibility of thinking and dwelling as opening to and for timing-spacing-thinging.

Subject-object dualism. We know already the importance Heidegger assigned to overcoming entrapment in this form of dualism. He referred to it as a prison that we carry around with us in all we do and said that his life's work was devoted to freeing us from such a prison.[3] From the start, in *Being and Time,* Dasein was not to be understood as a subject, and the beings of Dasein's world were not to be understood as objects. Neither could "being" be taken as an object of representation if the question of being were to be worked out so as to make being thinkable. The subject-object relation, as it plays out both metaphysically and epistemologically, would only be a hindrance in the attempt to genuinely raise the question of the meaning of being (GA 65: 252/CP 178). In terms of thinking in the crossing of the first and other beginning, the relationship of subject-object thinking to the history of metaphysics had to come to light. As soon as the

notion of ourselves as rational animals in a world of *substantial* presences
(beings) grounded on constant presence (being) took hold, the way was
opened to the long history of refining this understanding in terms of concepts representing other divisions such as subject-object (ID 32/96). I have
already pointed out that Dasein is a way of interpreting our ways of being-
in-the-world or our ways of being-on-the-way toward dwelling with things.
It should be clear, then, that to think of ourselves as subjects in a world
of objects over against us is also an *interpretation* of certain aspects of
human experience (GA 65: 488/CP 344). And just as seeing the emptiness
of be-ing does not make things disappear, neither does letting go of the
notion of ourselves as the subject of our experiences make us somehow
disappear. What it does is transform first our understanding and then (if
all goes well) our experience of ourselves in the midst of timing-spacing-
thinging (the main topic to be taken up later in this chapter).

When we see, once and for all, that "being" is not a being but an empty
representation that flattens the dynamic relationality of be-ing, then the
notions of "subject" and "object" are also decisively shaken. And in turn,
unless we can let go of our presuppositions concerning "subject" and
"object," we will be unable to let go of the habit of reifying beings. We will
find, then, that any attempt at understanding ourselves as opening for
timing-spacing-thinging—or any attempt to genuinely dwell with things—
will be blocked right from the start, because our modern notion of ourselves as *subject* has become "*the* refuge of those presuppositions" about
the nature of being, beings, and ourselves (GA 65: 443–44/CP 312–13; see
also GA 65: 261/CP 184). As the refuge and caretaker of these reifying assumptions, we are placed in a particular kind of relationship to being and to
beings, both of which take on the character of objects. But our place in the
midst of timing-placing-thinging is not thinkable in those terms. I mentioned already in chapter 2 that Heidegger never uses the German word
Relation to think this relationality but most often uses *Verhältnis. Relation*
says a relationship between entities. *Verhältnis* says "holding together," which
lends itself much better to saying the dynamic gathering of thinging and
the gathering of its showing as saying, the "relation of all relations" [*Verhältnis aller Verhältnisse*] (GA 12: 203/WL 107). So it is that Heidegger is quite
clear that to "talk of relation [*Relation*] of Dasein to be-ing obscures be-
ing and turns be-ing into something over-against" instead of placing us
in the midst of the enowning of be-ing; therefore, our situation in regard
to be-ing "is entirely incompatible with the subject-object relation" (GA 65:
254/CP 179–80; see also GA 65: 303, 455/CP 214, 319–20). So letting go of the

subject-object presupposition is necessary in order to begin to think be-
ing and to allow the possibility of being shifted with awareness into the
open play of timing-spacing-thinging. But it is not sufficient. For many
people the notions of subject and object, while they are undoubtedly in
play, are somewhat covert in their shaping of thinking and action. If you
ask Joe Average how he thinks of himself, it is likely that his response will
be couched in dualistic terms, but the word "subject" will most likely not
be part of his description. The philosophical concepts in play have been
taken over into general, popular thinking in more or less this way: Joe is
his mind and body, and as a human being he is distinctly different from
all other things, which are just objects.[4] So mind-body and self-other dualisms are also very powerful and must be confronted.

   Mind-body dualism. If we are not "beings," then there is no basis at all
to assume that any part of us is a being. That is the bottom line. But given
the powerful hold that mind-body dualism has on us, I will need to say a
bit more. Certainly, we *can* think of ourselves in terms of mental and physical processes. So in that sense the terms "mind" and "body" say something
meaningful and undeniable: we think, we eat, we walk and talk and sleep.
Here it is helpful to think in terms of interpretation, just as Heidegger suggested we do in understanding what he means by his ways of using the word
"Dasein." If we continue to speak of mind and body while trying to understand their (our) arising in the open play of timing-spacing-thinging,
we need to interpret them differently. We could say, with some insight,
that mind is something body does, but only if we also say that body is
something mind does. What basis is there, after all, for giving either one
precedence over the other?[5] In either case, with "mind" or "body" we are
abstracting out one aspect (or related complex) of our experience and giving it a name. Such abstracting pulls two ways. On the one hand, it does
name something we actually experience. On the other, if we make too much
of it, the abstraction serves to flatten the richness of our experience of and
as the dynamic relationality of timing-spacing-thinging.

   If we reify mind and body, then we will also tend to objectify the body.
With just a moment's reflection on what we receive daily by way of the
media we can see some of the results of this objectification as they play
out in advertising, pop psychology, diet fads of all kinds, and various forms
of moral hectoring. The current epidemic of young girls falling victim to
anorexia and bulimia is a direct result of mind-body dualism and the objectification of the body in connection with the culture's long-standing deval-
uation of women. In ways too numerous to count we are told over and

over that if only we—all of us—will think correctly and exercise some will-power, we can be beautiful, thin, smart, rich, and *good*. The reason this kind of information has the power it does in our society is that we assume that treating our bodies as objects somehow reflects reality. "It's just the way it is." Yes, it is indeed the way it "is" under enframing, in which not only the notion of being but also the dualisms based on its initial reification are taken to an extreme. But this "way it is" is not necessary; it is not the reflection of some deep, absolute reality. It further plays out the history of metaphysics. On the basis of the very first splitting of the idea being from beings, all these other divisions seem to be justified. And one can, of course, justify them conceptually with various arguments and then deal with all the philosophical problems that arise in the aftermath.

Instead, I suggest we raise a much more interesting question: How would we live and function if we were able to think differently, if we were able to let go of subject and object and let go of mind and body as notions that shape how we think of ourselves? That question calls on us to be open to other ways of interpreting the experiences that we now capture by those names. It calls on us to listen to saying, from out of our belonging to timing-spacing-thinging, in which we "are" dynamically open-dimensional. But it may be difficult to listen because of yet another form of dualism.

Self-other dualism. When Heidegger said that the thought of ourselves as subjects is the refuge of all dualistic assumptions, he caught hold of an important insight, one that hints at the underlying reason why these dualisms are so hard to release from their position of dominance. Dualistic thinking has, from the very beginning, given human beings (or some sub-set of humans, determined in various times and places by things such as race and gender and class) a very comforting sense of superiority. The divisions subject-object and mind-body have by no means been value-free notions that only pertain to the arcane domains of metaphysics and epistemology. Humans, the subjects having "mind" (and language, which is usually closely linked to any sense of what it is to "have" mind), are elevated over all other beings. All the way back to the origins of monotheism and on through the first beginning of Western philosophy down until now we find this as an underlying motif. Genesis 1:26–28 tells the humans that since they are made in the image of God (which is usually taken to include language, thought, and the capacity for spiritual and moral experience), they are to have *dominion* over all the creation, right on down to the last detail, to "every creeping thing that creeps upon the earth." Plato and Aristotle both distinguished us from the other animals in terms of our linguistic and intellectual capacity, and their doing so was by no means value-free. They also felt that men were more in possession of those capacities and therefore superior to women. Aristotle in particular was rather blunt about that. Then there is the notion of the hierarchical Great Chain of Being that developed in the Middle Ages. In the seventeenth century Descartes conceived all bodies—not just nonliving things but also animal bodies, including our own—as mere mechanisms. After defining the other animals as possessing no mind at all and therefore no sentience, he used these ideas to justify vivisection. These are all fairly well known developments. The reason I bring them up here is to remind us of the power that the dualism-based notion of human superiority has had and continues to have. Even after about thirty years of environmental philosophers' having argued against this kind of thinking, little has changed in terms of society's ways of interacting with the nonhuman realm. Why not? Because of the notion of human superiority (or male superiority, or race superiority, or class standing, or intellectual elitism, etc.). Dualistic thinking and the kinds of value judgments and behavior based on it have been dominant for so long that it will take more than philosophical argument and ethical theorizing to bring a change. If we are to even imagine letting go of these effects of self-other dualism thought in terms of human-other, we have to be able to first imagine letting go of self-other dualism on the personal level. That means, in terms of the thinking of be-ing, that we need to be able to think and imagine living as opening for timing-spacing-thinging. It means we need to be able to imagine making the leap to *dwelling*, to staying with things so as to nurture and care for them, freeing them *while and as freeing ourselves*.

I hope I have already indicated that willing ourselves to make different kinds of value judgments and refining our ethical thinking is the way to go. The question of whether or not one can "base an ethics" on Heidegger's thinking has received some attention over the last few years. I suppose one could, but it would mean violating what Heidegger's thinking is trying to accomplish. It would mean plundering Heidegger's work to achieve an end that conflicts with the most important insights he gives us. From very early in his thinking Heidegger tried to make clear that ethics was not something he was working toward. He tried to discourage people from misinterpreting the distinction between authentic and inauthentic Dasein in moral terms and said more than once that he had no intention of thinking in terms of ethics or values (GA 65: 60/CP 87; BW 251, 259). Why not?

In the first place, the thinking of the first and other beginning, if it is to persist in thinking be-ing and in opening up our situation in timing-spacing-thinging, cannot be reduced to any kind of dogma or doctrine. Heidegger is quite clear on why he rejects that. "When thinking comes to an end by slipping out of its element it replaces this loss by procuring a validity for itself as *technē*. . . . Philosophy becomes a technique for explaining from the highest causes[,] . . . [and] in competition with one another [philosophies] publicly offer themselves as 'isms' and try to offer more than the others" (GA 9: 317/P 242; see also GA 65: 7, 439/CP 6, 309). Philosophical theorizing, including the production of ethical theory, is rooted in the presuppositions of metaphysics. Ethics is rooted in metaphysics and its epistemology, and even thinking in terms of values and value judgments is rooted in subject-object dualism. "Precisely through the characterization of something as a 'value' what is so valued is robbed of its worth . . . admitted only as an object for man's estimation . . . a subjectivizing. . . . The bizarre effort to probe the objectivity of values does not know what it is doing" (GA 9: 349/P 265). The very act of making value judgments is rooted in the notion of our acting as subjects in relation to objects that, in and of themselves, have no "say" in the matter. So even assigning a very high value to something is an assumption that it has its value in relation to us. The fact that environmental ethicists, for example, think that they must offer a justification and argue on behalf of the notion of the intrinsic value of other kinds of beings or species or ecosystems is itself a tacit admission that, within the realm of ethical theory, traditional metaphysical and epistemological presuppositions hold sway. We, the human subjects, will decide whether or not to grant value to the objects with which ethics is attempting to concern itself. The very notion of "intrinsic value," in that framework, is rather perplexing, to say the least. No matter how convinced we are that the notion of human superiority is wrong-headed and destructive, until we can release our attachment to the underlying dualistic presuppositions, "value" can only be an admission of "no intrinsic value." Until we can let go of dualistically fixating on ourselves as subjects over against objects, as minds in charge of bodies (ours or others'), and as selves in relation to others, "ethics" will be both necessary and ineffective. It will seem necessary because of the requirement that harmful or offensive behavior be restrained. It will be ineffective because it does too little to actually change anything, fostering our sense of superiority along with the notion that if we can only figure things out well enough and exercise enough control over ourselves and others, we can solve the problems that confront us. But the very assumptions and attitudes that ethics and values thinking fosters are in large measure at the root of many of those problems (GA 65: 135, 139–42, 92–93, 441–42/CP 92–94, 97–99, 310–11).

So the bottom line is that to try to derive "an ethics" from the thinking of be-ing would be to expect both too much and too little. It would expect too much in looking for a way to solve problems and to expect this solution to come by way of the usual mechanics of theory production: concepts, principles, argumentation. The thinking of be-ing is, for this purpose, useless (GA 65: 396/CP 278). But the demand for an ethics also expects too little. The kind of transformation that becomes possible here runs much deeper than a willed change of attitude, a new set of values, or a different kind of ethics ever could. "With all the good intentions and all the ceaseless effort, these attempts are no more than makeshift patchwork, expedients for the moment. And why? Because the ideas of aims, purposes, and means, of effects and causes, from which all these attempts arise—because these ideas are from the start incapable of holding themselves open to what is" and thus fall far short of the kind of radical transformation opened up in the thinking of be-ing, the possibility of dwelling with things in mindfulness of our interrelationality in timing-spacing-thinging (WHD 64–55/WCT 66; GA 65: 37, 184–85/CP 26–27, 129). Furthermore, the very basis of ethical theory, its grounding on some definite idea of the nature of beings, is thoroughly shaken in the thinking of be-ing. "All calculating according to 'purposes' and 'values' stems from an entirely definite interpretation of beings. . . . [H]ereby the question of be-ing is not even intimated, let alone asked[,] . . . [resulting in] all noisy talk . . . without foundation and empty" (GA 65: 72/CP 50). The thinking of be-ing and its opening toward dwelling is not an attempt to solve our problems and aim at utopia through some kind of ethical-political planning. It is also not subject to the kinds of limits that pertain to such attempts, attempts that limit transformation to incremental change within predetermined bounds. We cannot predict what may come, but one thing is clear: a way of thinking that alters all our deepest presuppositions about ourselves and the nature of the world is going to have unimaginably far-reaching ramifications, *if* it can be thought and imagined and lived. I said earlier in this book, as the all-pervasive nature of dynamic relationality began to come to the fore, "Change one thing, and everything changes." If that "one thing" is the thought of be-ing, and the next thing is our understanding of ourselves, then everything else begins to follow. I hope it is clear that I do not mean "follow" here in the sense in which each premise in an argument follows from another. What follows from (and accompanies) the thinking of be-ing is a multifaceted shifting in which "all relationship to a being is transformed." Here the imagery of Indra's net is again helpful. I brought it into play at first to help explain the way that the joinings of guidewords work. Those guidewords, however, say and show something of the dynamic of the turnings in enowning, of timing-spacing-thinging. One way that Heidegger gives us to think the possibility of transformation is of *turning with* the turnings in enowning (GA 65: 407/CP 286). This transforming will not be subject to planning and prediction. On the contrary, it depends on being attuned to the dynamic of timing-spacing-thinging, being attuned to the reservedness that echoes being's withdrawal from any grasping attempt. We might well wonder, then, what comes next. We proceed by once again gathering ourselves to releasement toward things and openness to mystery.

RELEASEMENT TOWARD THINGS AS OPENNESS TO MYSTERY

If we can succeed to a significant extent in releasement toward things, letting go of the things that constitute calculative, metaphysical thinking and letting go of the most powerful notions that have emerged from that thinking, then "openness to mystery," already at work, can come even more to the fore. Releasement toward things is multifaceted, and, as it comes into play, it already begins to converge with openness to mystery. Why do I say that? Gathering up the "things" to be released and the sense of "releasing" enacted in each case will help answer that question. The phrase "releasement toward things" is ambiguous; that very ambiguity is part of its power in helping us engage the thinking of be-ing and open up what it might mean to dwell in timing-spacing-thinging. On the one hand, releasement means letting go of what blocks or hinders thinking and dwelling. On the other, releasement means releasing ourselves *toward* things, opening to them in a new way. What things are released in the first sense, and how are their releasements related to one another?

1. The traditional rules, norms, and expectations of what constitutes good thinking: conceptual grasping and fixing, method, theory, system.
2. The idea of "being" as something that is reified upon being conceptually lifted out and separated from beings.
3. The things that such traditional philosophizing begins from and works toward: *archē* and *telos*. Releasing the idea of being also releases the idea of its primary function: to serve as the ground of beings and of thinking.

Be-ing is ab-ground, and there is no *archē*, no first principle, to be found
in its thinking. With no *archē* there is no *telos*, no ultimate end or aim
that could somehow be attained in carrying out the thinking.

4. Therefore, the notion of "an ethics" is also released along with other
   kinds of theorizing (metaphysics, epistemology). Dwelling cannot be
   constrained within the frameworks of ethics.

5. The various interpretations of "beings" that are grounded on some concept of "being." This involves releasing such notions as substance (and
   the related philosophical notion of "accidents"), matter with its form,
   subject and object, and dualistically conceived mind and body.

6. The idea of ourselves as beings, conceived in any of the ways listed in
   point five.

7. And finally, the one that Heidegger himself gives when he first speaks of
   releasement toward things in "Memorial Address," letting go of our
   entrapped fascination with the products of techno-calculative thinking,
   including taking language as merely information or entertainment (GA
   12: 251/WL 130).

When Heidegger first suggested that releasement toward things was a step
toward being able to learn to think in a way that was not just calculative,
he described it as being able to say both "yes" and "no" to technical devices.
Releasement toward things does not mean flat rejection. In the case of the
things indicated in point seven they can be used or not used. What is released is the sense that they are somehow *necessary;* they become optional.
This "yes" and "no," indicating the optional character of what is released,
has a bearing on all the other items on my list, too, and we can examine
each of them in that light.

1. What would it mean to say "yes" to concepts and theory after first releasing them as constraints on thinking? It is probably the case, as I pointed
   out in chapter 2, that most language involves at least some grasping. Heidegger suggests that the guidewords for thinking be thought in terms
   not of the kind of grasping that concepts do but as in-grasping of the
   saying of things and language. In-grasping is not strictly intellectual but
   involves the *heart-mind;* it is a grasping that is able, at the same time,
   to let go of what is grasped, releasing any claim to having acquired some
   fixed and final "last word" on any matter. Another way that concepts
   must be taken into account in the thinking of the first and other beginning is due to the fact that they *do* say something (GA 12: 182/CP 87). In

their very grasping they say how something has been and is understood and—this is crucial—how the most powerful concepts of metaphysical thinking also conceal the withdrawing intimations of be-ing. So to carefully think with and through these concepts is essential (GA 65: 83–84/ CP 58).

2. I indicated already in chapter 1 that almost as soon as the thinking of the first and other beginning gets under way, opening up the awareness of the conceptual creation of reified being along with the emerging realization of abandonment of being, the very idea "being" is no longer compelling but optional. We may not know yet how to think of things, but we are no longer bound to think of them in the way handed down from the Greeks through the history of metaphysics. And in thinking "being" carefully we encounter, in several ways, intimations of be-ing.

3. The metaphysical concepts of *archē*, *telos*, and ground are three of those powerful concepts I spoke of just above, in point one. In holding these notions in question until thinking shifts into ab-ground, the notion of "ground" is not simply rejected but transformed, opening up Ur-ground as ground*ing*. The thinking of the first and other beginning is not vacuously free-floating and groundless but, rather, echoes its grounding in the utterly nonreifiable, always dynamic relationality of timing-spacing-thinging, already discussed. It also opens up the grounding that takes place in and as *Da*-sein, yet to be discussed. This is also one of the indications that releasement toward things will converge with openness to mystery.

4. Again, while the thinking of the first and other beginning releases any claim to ethical normativity, it examines the concepts put forward by ethics insofar as they say something of what concerns people and also carry traces of the unthought.

5. Letting go of the traditional concepts by which we have thought beings requires confronting what they say of how we experience ourselves and the world of things. In spite of the ways in which these concepts flatten experience they nevertheless are not simply "wrong." I will discuss "mind" and "body" in more detail in that regard below.

That is a careful look at releasement toward things in the first sense: letting go of things that block or hinder the thinking of be-ing.

But releasement toward things means more than just that. It also means releasing ourselves toward things in and as opening for them to say and show themselves. How? I began discussing this back in chapter 4 in the discussion of dwelling. Heidegger said in "Building Dwelling Thinking" that dwelling, as preserving the fourfold, actually means *staying with things.* So in my discussion of dwelling thus far I focused on clarifying this in a fairly straightforward, down-to-earth way. Staying with things means heeding them from out of our belonging with them in the same relational dynamic, timing-spacing-thinging. Because we belong we can listen to their saying (showing) and are then able to respond, to "say after" their saying, in language and in silence, in caring for and preserving their enowned, momentary uniqueness. We do this in genuine face-to-face encounter with things and in taking the time to think and act from within what arises in our heart-mind, the thanc, in such an encounter. My discussion of "Cooking Dwelling Thinking" was a first attempt to clarify what that might mean. There is, however, much more that can and should be said. Several threads in the thinking begin to converge and weave together here as we attempt to think "dwelling" more deeply: (1) releasement toward things, (2) openness to mystery and ineffability, (3) our "place" within timing-spacing-thinging, (4) grounding in ab-ground, and (5) the transforming of Dasein.

Embodying Mystery, Embodying Thinking

Just as it is easy to underestimate thinking it is easy to misunderstand and underestimate the power of openness to mystery, assuming that it might be some sort of vague, laissez-faire matter or that it might be foggy, imprecise, sentimental, or pointless in regard to our everyday lives. In chapter 4 I already gave some indication of how this is not at all pointless in that the *mystery* spoken of here is not "what withdraws" understood in itself, as if it were something "out there." The mystery is timing-spacing-thinging itself, which is *never* apart from things or apart from us. Thinking this mystery in terms of multifaceted emptiness and fullness opens the way for transforming all of our relations to what appears in us and around us.

From the start the *Da*, the t\here of Dasein, means *opening.* It never means "the open" thought as a reifiable void. So when we read in Heidegger's later work things like "Dasein shifts into openness" it is important not to begin to reify this as if it were an object that we could, as subjects, come to know. Dasein is more so to be thought in terms of our way of being open, so as we think it we remain aware of its dynamic character. This opening is not something that we can, through an act of will, open up. It is, instead, the open play of timing-spacing-thinging in which we always already find ourselves but only now first with clear awareness (GA 65: 237,

259–60, 296, 304/CP 167, 182–83, 209, 296). As we attempt to think more deeply into the t\here many guidewords come into play, resonating with one another. "The t\here is the open between that lights up and shelters— between earth and world . . . and thus the site for the most intimate belong-ingness, and thus the ground for the 'to oneself,' the self and selfhood. . . . With the grounding of Da-*sein* all relationship to a being is transformed, and the truth of be-ing is first experienced" (GA 65: 321–22/CP 226). Da-sein is here said to be open, between, and grounding, in which all our rela-tions to a being (any being, including us) are transformed. Of the facets brought into play in this saying I have said more so far of opening and transforming in relation to Dasein than I have of the "between" and ground-ing. The opening that is the t\here of Dasein is none other than the open-ing cleared in timing-spacing-thinging. As this opening is also said as the turnings in enowning, it (as such) opens ways for manifold transformative possibilities. Everything is constantly changing; openness and awareness enable us to go with rather than counter to this constant turning (e.g., as we have done in our grasping and reifying of being and beings and our-selves). How it is that *all relations* to things are transformed will only emerge over the course of the discussion.

Dasein is said to be "the between" for the strife of earth and world, that is, for thinging. This calls for a bit more thought. It is one thing to say that Dasein emerges within the same play of timing-spacing-thinging in which things emerge and another to say that Dasein is the between, the "site" for this gathering. What does that mean, that Dasein is the between? First, let me say what can be ruled out as misleading. Dasein is not the only way to say "the between." The between here is, as I said, none other than the opening-clearing in which the gathering of things is enowned. It is radi-ant emptiness that is the very fullness of thinging. Therefore, Dasein is not the owner or creator of the between but, as Heidegger puts it, its guardian (GA 65: 484/CP 341). We also need to keep in mind that this thinking is not linear, working with cause and effect, but moves in the resonance of joinings. So it is not the case that Dasein is somehow there (in time and space metaphysically conceived), between things, to take over guardianship of be-ing and the things that emerge in it. Instead, "the between of Da-sein . . . [does not] build a bridge between be-ing (beingness) and beings—as if there were two riverbanks needing to be bridged—but by simultane-ously transforming be-ing and beings in their simultaneity. Rather than possessing an already established standpoint, the leap into the between first of all lets Da-sein spring forth" (GA 65: 14/CP 11; see also GA 65: 230–31,

387/CP 163, 271). What does it mean, then, to say that Dasein is, as the be-tween for be-ing, the *guardian* and *grounding* of be-ing's holding-sway in the enowning of things? The key, I think, is to call on the uniqueness of Dasein as Dasein, namely, our capacity for mindful, knowing awareness, in the thinking of be-ing. The thinking of be-ing, as ab-ground, is never separable from timing-spacing-thinging. There is no being, no ground, out there somewhere, transcendent to things (timing-spacing-thinging). But this also means, insofar as we also arise in the midst of timing-spacing-thinging, that there is no ground transcendent to us. And our particular uniqueness, as Dasein, is to be open to ab-ground, to think be-ing, to come to knowing awareness (which does not mean conceptual grasping) of timing-spacing-thinging. What arises then is a mutual, simultaneous grounding of Dasein and things in which—and only then—we are *Da-seyn*, the t\here for be-ing's timing-spacing-enowning, owning this as what is ownmost to ourselves.

This owning of the enowning of the t\here first allows us to turn with the turnings in enowning and engage the thinking of be-ing in such a way that we become open to its full range of transformative possibilities; that is, we are capable of thinking as dwelling. To think this in terms of ground-ing, enowning, be-ing, and timing-spacing, though necessary, can tend to lead away from where we actually dwell, that is, with things, if we let it. So as we carefully move with the thinking of how our transforming relations to things may come to awareness I will offer reminders: this is speaking of our everyday relations with things, and we will come around to that again and again in the course of thinking. After all, not only is the enowning of timing-spacing-thinging deeply ineffable, but the things themselves (includ-ing us) that are gathered in it are also extraordinary and ineffable in their very ordinariness. Be-ing, enowning, timing-spacing-thinging, things, us, saying-showing: every last one, by any name, is a magical display of radi-ant, ineffable emptiness. Most remarkable of all is that we can think this! We can, in coming to knowing awareness, learn to dwell with things in such a way as to be the caretakers of be-ing. We can do this because it is ours to belong to timing-spacing-thinging with the capacity for *hearing* what it says to us in each unique, unrepeatable moment (GA 65: 251–52, 380–82, 407–8/CP 17–18, 265–67, 286–87).

I used as the epigraph for the entire book this thought of Heidegger's: "To think is above all else to listen" (GA 12: 170/WL 76). It should now be clear why I did so. Our capacity to hear and heed what things, in timing-spacing-thinging, show to us is central to every aspect of thinking with

Heidegger and thinking after Heidegger. It is the core of the possibility of transformation that arises in the thinking of be-ing. The between of *Da-seyn*, as discussed above, is the place of opening to the saying of things. "Saying is showing. . . . Saying is in no way the linguistic expression added to the phenomena after they have appeared, rather all radiant appearances and all fading away is grounded in the showing saying. . . . Saying is the gathering that joins all appearances in the in itself manifold showing which everywhere lets all that is shown abide within itself" (GA 12: 246/WL 126; see also GA 65: 342/CP 240). Our place within this shimmering, res-onating saying of timing-spacing-thinging is what enables both thinking and dwelling as co-responding, as saying after saying. I already said much of this in chapter 2. However, now, having thought timing-spacing-thinging as radiant emptiness and worked carefully through all the things that need to be released to proceed with this thinking-dwelling, it becomes possible to begin to open up with more depth and clarity *how* one might *thought-fully dwell* with things without falling back into any form of reification.

Having let go of the limiting notions of subject and object as well as the dualism of mind and body, it becomes possible to consider that we *em-body* timing-spacing-thinging. Therefore, we must *embody* the thinking of timing-spacing-dwelling. Everything said from chapter 3 up to now points in that direction. There is no "mind" separate from "body." There is no "I" separate from the opening-grounding-between of timing-spacing-thinging. That by no means implies that we somehow disappear or dwindle into insignificance. On the contrary, it enables us first of all to open to the full richness of our capacity to think and dwell. Thinking is not the activity of some disembodied or accidentally or temporarily embodied mind but arises as the thanc, the heart-mind. This fully embodied thinking, in con-tiguity and intertwining with things, engages with things in face-to-face encounter, gathering and bringing to the encounter all our memories and gratitude and openness, letting them stand "before us and giv[ing] our heart and mind to the 'being' of particular beings" because their gathering is none other than who and what we also are (WHD 137/WCT 226; see also WHD 91–94, 157/WCT 139–44). To be willing and able to open to things in this way is to be open to the possibility of undergoing radical transfor-mation and, in fact, to have undergone some significant transformation already. Embodying thinking is a far cry from thinking of ourselves as the cobbled-together "rational animal" (GA 65: 3/CP 3; see also GA 65: 84, 338/ CP 58, 237).

That thinking is embodied emphasizes in yet another way its necessarily being open to mystery. We may from time to time delude ourselves in the manner of Descartes that our "minds" are crystal clear to us, but our bodies are quite another matter. They are the deep well of all our past, all our memories, and all that we *do not remember* and, indeed, of what we can never actually quite know as well. They carry on without much rational control on our part, and aren't we grateful that they do when we stop and think about it? I say "they" as if to separate the body from me, the one thinking of it. That is nothing more than an artifact of our metaphysically constructed grammar.

Any discussion of what Heidegger means when he uses the word "thinking" eventually gets around to the way that thinking is drawn by and pulled along by the intimations of what continually withdraws from thought. I have discussed that to some extent already, beginning in chapter 2. Be-ing, enowning, saying, timing-spacing-thinging all withdraw in that they are empty, they are not beings, but nameable-though-inconceivable facets of the gathering of things into their appearing. One of the most often cited texts that discusses this being-drawn by what withdraws also says a bit more about why we are drawn to think what keeps eluding our grasp. We are drawn, it says, because we (1) are related to what withdraws in that we (2) bear its "stamp" in what is ownmost to us as humans, and, as such, (3) we ourselves are pointers toward what withdraws (WHD 5, 51–52/WCT 8, 17; see also GA 65: 245/CP 173). This is much less abstract and more understandable once we realize that we *embody* "what withdraws." Our own bodies, arising in and as opening for the relational dynamic of timing-spacing-thinging, are themselves as ineffable as anything we might call by some more arcane name like "enowning" or "be-ing." Our own bodies are themselves ineffable be-ing, enowning, timing-spacing-thinging. My own body, my body-mind, my fully embodied heart-mind, by whatever name, is in itself an Indra's net of complex mirroring and resonating. Just as I would assert that without our being embedded in the world of things we could not imagine thinking, so also I say that without our bodies I cannot see how anything at all that could be called *thinking* would be possible. Calculation, perhaps, on the order of a computer's work, but not thinking. Of course, a computer has neither body nor mind and is, therefore, fairly irrelevant to this discussion, except insofar as one of the concerns expressed by Heidegger is that, if we become enthralled with techno-calculative thinking, we lose the chance to do any other kind of thinking. We will have lost what is ours as humans: the capacity to think and to be more than computing machines.

It is our fully embodied heart-mind that, in the thinking of the first and
other beginning, is moved by and attuned to the shock of abandonment
of being, deep awe at the intimations of mystery in the withdrawing of
be-ing, and reservedness that, in our thinking and saying, echoes that mysterious withdrawing. It is our fully embodied heart-mind that in hearing
and heeding these things as well as the saying of things in timing-spacing-
thinging is able to corespond and say after saying, that is, to bring the
thinking to language and to dwell with things. It is the fully embodied heart-
mind that *thinks*. "From of old we move *within* a projecting-opening of
be-ing, without this projecting-opening ever becoming experienceable *as*
projecting-opening" (GA 65: 449/CP 316). *We* move within the throwing
open of be-ing, not as minds or only as bodies somehow passively affected
by the history of metaphysical thinking but as fully embodied "beings" that
eat, sleep, work, play, are moved with deep feelings, think, and fail to think.
And it is as such that we inquire into the meaning of being, become aware
of the way the Greeks first began to conceive of being separated from beings,
and learn to examine and release the concepts that arose from that creative
opening, letting them become optional so that our thinking can now open
to be-ing. Any knowing awareness of enowning that arises in us arises into
our fully embodied heart-mind, the thanc, which is attuned now, finally, to
the saying of things in the mystery of the withdrawing of timing-spacing-
thinging.

It is only our fully embodied heart-mind that could ever possibly hear
the saying of things. How could a "mind" ever come face-to-face with a
tree, a woodchuck, the dragonfly skimming over the pond's surface, or the
handful of loam that is cradled in the hand with the thought that "now
is the time to plant the peas; the soil is just right." As Heidegger puts it,
"The gathering of what is next to us here never means an after-the-fact col-
lection of what basically exists but the tidings that overtake all our doings,
the tidings of what we are committed to beforehand by being human beings.
Only because we are by nature gathered in contiguity can we remain con-
centrated on what is at once present and past and to come," which is the
gathering of what is to be thought and what may be thought in the thanc
(WHD 157–58/WCT 144–45). It would be a very odd sort of position to take
to think that we, as humans, as those who can think in this way, are contiguous only with things but not within our own body-heart-mind.

Unlike the main thrust of Western metaphysics and epistemology, which
takes the body as a hindrance and obstacle to the acquisition of truth,
of knowledge of the real, of the presumed certainty of clear and distinct

concepts, and so on, what opens up in the thinking of be-ing is the realization that with no embodiment there could be no thinking. And even the word "embodiment" here says much more than the tradition has ever allowed. It means not only our body-heart-mind but also the entire relational dynamic on which it arises and depends. And it is only within this full embodiment that thinking can take place. Instead of embodiment putting up a roadblock to thinking, *it is embodiment in this full sense that clears and opens the way for thinking.*

> Embodiment is . . .
> *Da*-sein's manifesting, which is . . .
> Opening for and as . . .
> Timing-spacing-thinging . . .
> Gathering in contiguity with things . . .
> All of which are in ongoing dynamic relational play . . .
> Saying and thus showing forth something in each thing . . .
> Clearing and making way for thinking . . .
> Which, fully embodied and enacted, . . .
> Is none other than dwelling . . .
> In the clarity that is the radiant emptiness of be-ing . . .
> Through staying with things, . . .
> Freeing them in their momentary uniqueness . . .
> And only thus freeing ourselves . . .
> For the knowing awareness . . .
> Of each unrepeatable moment . . .
> Of the radiant emptiness of be-ing.

Dwelling: Staying with Opening

We are, according to Heidegger, opening in the way I have just said. This opening is thoroughly be-thinged, that is, conditioned (*be-dingt*). We are what we are only in our ongoing, ever-changing dynamic relationality with all other things. This is our dwelling, this *staying* as opening to and with things. It is only this opening that enacts the freedom that allows the multifaceted transformations that begin to emerge in the thinking of be-ing. The notion of freedom as the transcendence of embodiment and of our entwining with the world of things is nothing more than an imaginary freedom that bears no relation to our actual experience. Dwelling as opening, as staying with things, enacts the radical—from the ab-ground root of radiantly empty be-ing—transformation of all our relations with things, with language, with ourselves and others (GA 65: 248/CP 175). It is enacted as *freeing*, both as freeing ourselves and as freeing the things that we encounter. Already, letting go of reification and accepting the awareness of the radiantly fluid emptiness of all things is freeing. It frees us from dualistic fixations that entrap us into predefined ways of thinking and acting. It frees us to begin to see (and hear, and touch, and think, and all of these at once) the extraordinary and deeply mysterious radiance of empty be-ing in each ordinary thing. It frees us into the kind of wonder at things that allows us to stop and *think* about our relations with them and to take the time to let our enowning and theirs—an enowning that cannot be thought of in terms of "one" or "two"—to move together. It frees us for releasement *toward* things in openness and attuning to their (and our) mystery. Heidegger says that it is this freeing that allows us to dwell with things such that we set them free as well (GA 65: 480–83/CP 338–40; TB 65; GA 7: 150–51/ PLT 148–49). What does that mean?

It does not mean simply pulling back and letting whatever happens happen or simply refraining from all action in regard to things. Again, that could only be an imaginary sort of freedom or freeing, bearing no relation to our experience of things. We are already thoroughly entwined with things, mutually shaping one another, with no way out short of death (a rather unappealing notion of freedom). Words that tend to come into play and join with one another when Heidegger discusses this matter are opening, clearing, freeing, lighting, lightening, gathering, sparing-and-preserving (one word: *Schonen*), and, of course, thinking (TB 64–66; GA 7: 150–64/PLT 148–61). What does this joining say to us? In the first place, it evokes everything that has already been said of our situation in regard to the relational dynamic of timing-spacing-thinging as briefly recapitulated just above. This is the site of our dwelling, moment by moment. These are the moments when the openly radiant play of timing-spacing-thinging breaks through and arises into our fully embodied awareness in face-to-face encounter with things, not "things in general" but *this* thing. This fig, this stapler, this blade of grass, this hand held in front of my face, or these hands holding the injured goldfinch, or this ladybug eating an aphid, or this bolt of lightning, or this sip of gin and tonic sliding coolly down my throat. In Heidegger's discussion of space in "Building Dwelling Thinking" he helped us to think clearly through to the insight that we are not just encapsulated bodies. We are not just "physical objects" housing a mind, as my discussion of the dualisms released by this thinking made clear. This also must open up our understanding of things. We already know that things are their own gathering of timing-spacing-thinging, so they cannot be thought of as encapsulated objects. "Thing" is not at all synonymous with "object." The scope of what thinking can call "thing" comes to awareness only in encounters with things themselves. The conceptual determinations of "part and whole" or "one and many" cannot be overlaid on the thinging of the thing. It cannot be predetermined or predefined. That is one way that we free things in learning to dwell with them: we learn to let them say themselves rather than assigning our own designations to them ahead of time.

This also frees us in that our own responses to things are not to be conceptually predetermined. This leads me into a question that has to be taken up at some point. One of the main ways our responses to things and, even more so, to each other get predetermined is by way of ethics. I have already said that ethics is not dwelling and that it is, in fact, a hindrance to dwelling. But then how are we to dwell, how are we to live well and thoughtfully, if not within some ethical framework that can guide our thoughts and actions? In the first place, there is no substitute for *thinking*, and we must do this thinking ourselves. That does not imply, of course, that it happens in a vacuum, ex nihilo. If that were the case, I would not be writing a book called *Thinking after Heidegger*. On the other hand, Heidegger himself was quite clear, as I have already pointed out here and there, that (1) his thinking is not to be taken as any kind of doctrine and (2) each person must do her own thinking in her own situation, each time. As he put it, if all we intend to do is parrot Heidegger, then we would be better off to burn our "notes, however precise they may be—and the sooner the better" (WHD 160/WCT 158). The operative word here, in my title, is the ambiguous *after*. After: in the manner of, according to. After: later, going beyond. However, since it is Heidegger whom I am thinking after, the two senses converge. Thinking in the manner of Heidegger, in the way opened up by Heidegger, is to think not according to Heidegger but *according to the matter for thinking*, which always goes beyond any one saying attempt, no matter how brilliant it may be. The ineffability of the radiant emptiness of be-ing always refuses to be captured in words. Thus the most careful and insightful and co-responding thinking will always open onto more; it will always have something to say later as it continues to engage the matter. This matter, be-ing, is also said in this way: timing-spacing-thinging. Be-ing is inseparable from us, from things. So the matter for thinking, what calls for thinking, is to be engaged as it comes to meet us in language and as it comes to meet us in our encounters with things.

In both cases, having let go of a predetermined response, our co-responding to the saying and showing in language and of things must be a *spontaneous response*. And there you have it: my response to the question of how we enact dwelling in the absence of ethics. This spontaneity arises in and accords with the simultaneity of timing-spacing-thinging, which is our t\here, whether or not we are aware of it. The point of thinking is to become aware of it. The simultaneity of timing-spacing-thinging is, then, the simultaneity of radiant emptiness and lucidity.[6] Therefore, "spontaneous response" is nothing at all like any mere thoughtless, knee-jerk, "on the spot" reactivity. It is, in fact, quite the opposite, as it arises in each case from our fully embodied heart-mind, the thanc. In chapter 4 I described, in a preliminary way, how thinking arises in the responsiveness of our heart's core, the heart-mind, the thanc. What I am saying here is that this thought carries over into the question of our responses and actions in relation to things. How we respond and relate to things also can emerge spontaneously from the thanc, without rule, or doctrine, or ethical normativity having any role to play.

Some philosophically oriented individuals will hear "spontaneous action from the heart's core" as emotivism. That, of course, is far off the mark. "Heart-mind" is not reducible to "emotion" any more than it is to "intellect." The thanc is neither one nor the other pole of any such duality, nor is it the linkup of two separate components, dualistically conceived. *Thanc is one way of saying how we embody and enact the energy of timing-spacing-thinging in our thinking and dwelling*. In particular, it says our *being moved* to respond to saying, the saying of things and the saying that speaks to us in language. We can be moved in this way because it is ours to be t\here, inseparably open in and to timing-spacing-thinging. We begin to come to awareness of ourselves as open and cleared for the lighting up of things when we ourselves lighten up, releasing our tightly held sense of our own separateness, and pay attention. This allows the inseparability of radiant emptiness (be-ing, timing-spacing-thinging) and our own lucidity to converge and move us to responsive thinking and doing. Some may take "responsive" to mean "passive," which it does not, any more than it means "active," thought in opposition to passivity. By now it should be clear that there is *no* dualistic dichotomy that is going to be applicable to thinking and dwelling. This responsiveness is our dwelling in the dynamic between of what calls for thinking and our fully embodied belonging to it (GA 65: 56–57/CP 39–40). As one kind of illustration, Heidegger gives us the example of the skilled woodworker who, instead of doggedly cutting the wood to a plan purchased at the lumberyard, "learns to respond to the shapes in the wood." Without a great deal of thought and experience this kind of responsiveness will be very difficult. Yet in play with thought and skill it can happen with graceful spontaneity that allows the wood to emerge into something of beauty (WHD 50/WCT 14–15). It is the uniformity of enframing that inculcates passivity; thinking and dwelling are, by contrast, dynamically and freely responsive.

I have put so much emphasis on things, from chapter 4 on, that language has almost receded into the background as an explicit matter for thinking. An understanding of the working of language has, of course, been in play throughout. Saying, as what is the ownmost heart's core of language as well as the showing forth of things themselves, is the relationality that enables us to encounter things and to think. The sheer energy and power of language to shape us is a matter of incredible importance if we are to aspire to genuinely think and dwell. I have emphasized the kinds of things that limit and constrain us and therefore have to be released in the attempt to think be-ing. They have to be released again and again, since they are somewhat like cockroaches, scurrying around in the dark, scuttling under cover as soon as we try to turn on the light of thought to examine, trap, and release them.[7] However, the energy of language is also opening and freeing, as it attunes our thinking to the dynamic of timing-spacing-thinging. Our reading a passage of careful thinking or a poem or even a telling word can enter into the thanc and move us to carrying forward with thinking and dwelling in a fresh and creative way.[8] Thus language itself can at times lead us into the midst of timing-spacing-thinging, offering deep insight into dwelling with things. This is the work of saying.

Saying, the heart's core of language, does its showing work in many ways, in word and wordlessly, in sound and stillness. In our busy, noisy world, finding the simple stillness to heed saying is a challenge. I am speaking not only of the need to slow down and pay attention but also of the demands of actual *noise*. After September 11, 2001, in those few days when the commercial airlines did not run flights, it was astonishing how even the absence of that one kind of pervasive noise opened up a fresh awareness of the comparative quiet of the woods around our house. As I sit here writing this, however, I hear a lawnmower, crickets screeching, dogs barking, a truck revving its engines over on the next hill, a neighbor's son firing his 12-gauge shotgun (I hope all the doves are here, at the bird feeders), a dog whining in fear at the sound of the gunfire, the clicking of the computer's keys as I type these words, and, as I pause, the faint sound of the breath moving

into and out of my nose. I am prone to react to all this noise with annoy-
ance: no wonder it is so hard to think! But if, instead, I attempt to think
the noise, to call it into question, something else might arise. I cannot simply aim at getting rid of much of this noise, appealing though the notion
may be: cast all this noise into nonbeing! Instead, can I *think* with what is
emerging here? Noise . . . not-being. Silence. Two Tibetan words come into
my awareness: *chem* (noise) and *chemmeba* (silence, stillness, not-noise).
The ending of the word for "silence" reminds me of *medpa*, the word that,
though it literally means not-being, can also be translated as "ineffable,"
as I mentioned in chapter 5's discussion of emptiness. Emptiness is not
simply the cancellation of being but the inconceivable radiance of timing-
spacing-thinging. We are so oriented toward the sense that dominates in
us, vision, that we tend to overlook the play of sound in timing-spacing-
thinging. Thinking *chem* and *medpa* together evokes this thought: the ineffability of noise.[9] Put that way, it sounds rather odd. How about the
ineffability of sound? This hints at something much deeper. "Saying, as
the way-making movement of the world's fourfold, gathers all things up
into the nearness of face-to-face encounter, and does so soundlessly, as quietly as time times, as space spaces, as quietly as the play of time-space is
enacted. The soundless gathering call, by which saying moves the world-
relation on its way, we call the ringing of stillness" (GA 12: 203–4/WL 108;
see also GA 65: 510/CP 358–59). The ringing of stillness: the silence that
opens the way for all sounding, the stilling that yields place for all moving,
all arising into appearance and saying of things. How to move into and with
this ringing stilling is a question for further thought on other occasions.
Here, we see how lightening up and paying attention to something that at
first seems to hinder thinking may open the way to a thoughtful insight
into the arising of things.

Sometimes, deep thinking can arise in our encountering a thought that
seems so alien that we at first cannot understand it. As an example I want
to follow a train of thought recorded by the great thirteenth-century Japanese philosopher Dōgen. He begins this way: "You should reflect on the
moment when you see the water of the ten directions. This is not just
studying the moment when humans . . . see water; this is studying the
moment when water sees water. Because water has the practice-realization
of water, water speaks of water. This is a complete understanding. You
should go forward and backward and leap beyond the vital path where
other fathoms other."[10] In our usual, habitual way of thinking it is absurd
to say "water sees water" or "water speaks of water." Having thought with

and after Heidegger to this point, we can understand the saying of water: it shows itself as water, as this water, here and now. How is it that when we see water and, in so doing, listen to the saying of water that "water sees water"? Dōgen's saying-after the saying of water in his face-to-face encounter with water reminds us that in the play of radiantly empty timing-spacing-thinging we enter into the water, and the water enters into us, in mutual gathering. In our lucid awareness of this gathering, if we can move with the resonating back-and-forth of thinging and make the leap out of dualistic fixation, then water sees water. This, Dōgen says, is a complete understanding. Complete but not final, as this very water is itself ineffable. Later in the same essay Dōgen goes on to say, "There is a world in water. . . . There is a world of sentient beings in a blade of grass."[11] This serves to reinforce the basic understanding of the dynamic relationality of timing-spacing-thinging and prepares us to attempt to understand his concluding thought. "Mountains are mountains, waters are waters. These words do not mean mountains are mountains; they mean mountains are mountains. Therefore investigate mountains thoroughly. When you investigate mountains thoroughly, this is the work of the mountains."[12] Mountains are mountains: they are the uniquely momentary gathering of each mountain. But mountains are not "mountains." Mountains, rivers, a blade of grass, this hand: each is ineffable, beyond being captured and grasped by any concept. Mountains are the radiantly empty timing-spacing-thinging of mountains. If we can see and hear and think this ineffable mountain as mountain, not as "mountain," then this is the work of the mountains. It is not just our own human initiative and creation in language. It is our coresponding saying-after the mountain's own saying.

But these days, if we are paying attention to mountains and rivers, we are likely to be confronted with the agony of their forests and cougars and fishes. Wild things do not fare well under the rule of techno-calculative enframing. And it is not just wild things that are endangered. Wild things stand outside the framework of standing reserve and are in danger in the resulting perception of their uselessness. Whatever is brought into a slot in standing reserve is also suffering. It makes no difference: human, tiger, mussel, Tennessee coneflower, condor, or passenger pigeon. Anyone who is oblivious to the distress of the natural world is simply not paying attention. From the mid-1930s forward Heidegger was deeply concerned about this matter, as I discussed in chapters 1 and 2. He does not whine or complain or engage in polemics. Instead, he persistently and patiently *thinks* what is going on as well as its roots in the first beginning of Western thinking (which, he is clear, is reinforced by monotheistic thinking that devalues earth in its own way). The upshot is that if the situation does not turn around, we are left contemplating the destruction of our planetary ecosystems and their ability to continue to support the intertwining web of life. Heidegger poses this series of questions for our consideration. "Why does the earth keep silent in this destruction? . . . Must nature be surrendered and abandoned to machination? Are we still capable of seeking the earth anew?" (GA 65: 277–78/CP 195). Earth keeps silent because *silent* saying is what is ownmost to earth as earth. But this silence is nevertheless *saying;* it has much to tell us *if we will only listen.* And we must listen with much more than our ears. This silent saying of earth calls on us to pay attention with our full embodiment and to let what it says enter our heart's core, the thanc.

This is, especially now, difficult. It is difficult not only in light of all the hindrances both within ourselves and in the world around us. It is very nearly unbearable for quite another reason. We begin to listen, and what we hear from the earth, from the plants and animals and waters, is deep distress and, all too often, death agony. This is compounded when we also open to our fellow humans and pay attention to what their lives say to us. I am thinking of the children of Iraq, of the women of Somalia and Afghanistan, of any father grieving over his child dead of starvation, of the people suffering and dying from cancers brought on due to corporate greed and carelessness with various kinds of toxic waste products. How are we to stay *open* in the face of all this misery? We cannot simply ignore it, because to ignore it would be to ignore the context and the roots from which it grows. But it is so huge and so painful. That pain, insofar as we feel it and heed it, has something to say to us. It tells us that in spite of it all our heartmind is not numb and dead and unresponsive. It enables us to begin to respond to Heidegger's question of whether we are still capable of seeking the earth and ourselves and each other anew (which is the same as asking whether we are capable of learning to think and to dwell) with a tentative "yes." As Joanna Macy puts it, "We are capable of suffering with our world, and that is the true meaning of compassion. It enables us to recognize our profound interconnectedness with all beings. Don't ever apologize for crying for the trees burning in the Amazon or over the waters polluted from mines in the Rockies. Don't apologize for the sorrow, grief, and rage you feel. It is a measure of your humanity."[13]

It opens to us the possibility of owning our capacity to be the opening for the saying of things and to dwell with them in spontaneous responsiveness. But it is quite clear that in doing that we are not able to simply turn away from enframing and either ignore or reject it. It does not allow us either option. We have, along with Heidegger, taken the first necessary step in thinking clearly about the origins of enframing as the culmination of a long history of metaphysical thinking. This at least allows us to see that there is no ultimate necessity in how we now think and live. It loosens the all-encompassing character of the hold of techno-calculative thinking on us. But what then? To begin to move in other directions in thinking and dwelling is a long and unpredictable process, not subject to calculative planning or ethical regulation. Heidegger offers a thought that, together with what Macy said, provides some insight into the character of our attempt to think and dwell with things and open up new possibilities while under the domination of techno-calculative enframing. "This restoring surmounting is similar to what happens when, in the human realm, one gets over grief or pain" (QT 39). As many of us can testify from experience, one does not ever just "get over" grief. It is a long-drawn-out process, with many ups and downs. This is, I think, what Heidegger is suggesting here. Once we realize the nature of the thinking that responds to the call of the saying of timing-spacing-thinging and let go of all the modes of reification and calculation that block it, the way is open to attempt dwelling. But neither dwelling with things nor its thinking unfolds like a well-signed and freshly paved highway. It is more like a mountainous hiking path with side trails and switchbacks and at times even dead ends in gullies. It takes work. And to think "work" without "plan" or "aim" is yet another radical difference in what is being attempted.

I say "work" and yet I also said "lighten up" and pay attention. It takes both and more yet. In between the lightening and the effort is our opening to being attuned to the radiant emptiness of timing-spacing-thinging. We hold back and pause, and there it is: the ineffability of all sounding and resounding, speaking to us in the deep stillness at the heart of all sound, resounding in our heart's core, the thanc. We look around and see that, in spite of all the misery and grieving, there is so much beauty that is trying to say something to us. It gives us pause. It makes us cautious, not too quick to assume that we always know what to do. We pause and listen and respond to what is said, not just to what we preconceive. This reservedness, the attuning to a new beginning for thinking and dwelling, opens the way to a more careful, heedful dwelling. All of this talk of caution and care and reserve can sound constraining, but, thought well, it opens onto the possibility of freeing ourselves and the things of our world.

I think back to my long-ago encounter with the mother woodchuck.

Would I do anything differently if that happened today? No, I would not. Somehow, without any preconception, I responded spontaneously by lightening up and paying attention. I was still, not just in the obvious bodily way but in my heart-mind. I was not filled with busy thoughts and questions and worries. The usual questions that jump up when one is approached by a "wild" animal in an unusual way did not even cross my mind. ("Is she sick, is she rabid, does she have distemper, should I go for the shotgun?") For just that moment—a moment whose length was decided by her—I stayed with her. I did not, back then, know how to understand what she said, though it obviously resonated with the experience a few days previous to that, of being the eyes of the earth. I wondered, over the years, why something like that never happened again. It did, only I hardly noticed. I was looking for something extraordinary, something unusual. I was not looking or listening for the extraordinary that is with us always, in every ordinary thing we encounter, if only we can lighten up and pay attention enough to notice.

Right now I will turn off the computer, find a comfortable place at the edge of the pond, and wait for whoever comes today with something to say: the darner dragonflies, the green heron, the bullfrogs, the stones between the rushes. Maybe one of the dogs will keep me company, maybe not. I care deeply about all the misery in the world and hope to do my part in helping us all get over this long grief. But that does not stand in opposition to sitting still and just letting go of all of it in openness to the radiant emptiness of timing-spacing-thinging that is the extraordinary heart of each ordinary thing. This stilling that can open and attend to what comes to encounter us is essential for thinking, for dwelling. Only in such stilling can we open and hear the saying that, in each unique thing, intimates the mystery of the energy and power of its enowning as timing-spacing-thinging. This is the "soft power" that Heidegger says, in a brief and beautiful piece called "The Country Path," can outlast and overcome even the most brutal manifestations of techno-calculative thinking.[14] How so? In releasing the compulsive urge to take charge and force a calculated change we open to the mystery of our enowning as openness, as fully embodied lucid awareness of our place in the midst of all that arises. This enables us, as Heidegger puts it, to turn with the turnings in enowning and thus opens the way to spontaneously respond to things and situations as they show themselves. This is in no way a retreat from responsibility for the way things are going in our world. On the contrary, this is the enabling of respons-ability, the enabling of a response that is not trapped or channeled by enframing. Only then, in our fully embodied response, from out of the thanc, can we and things be mutually freed. "Only then is fulfilled the full uniqueness of enowning and of all momentariness of Da-sein that is allotted to uniqueness. Only then is the deepest joy freed from its ground, as the *creating* which by the most reticent reservedness is protected from degenerating into a sheer and insatiable driving around in blind urges" (GA 65: 249/CP 175).

This is the deepest and most powerful of grass-roots movements. As Dōgen said, there is a world of sentient beings in a blade of grass. Each of us is only one of them. But we are never isolated in our grief over the devastation of the earth that is unfolding all around us. Heidegger once asked, Why does the earth keep silent in this destruction? She is not silent. She speaks powerfully through the voices of each lake and frog and thistle, every forest and boulder. It falls to us to listen and to join our voices to theirs.

NOTES

1. OPENING A WAY

1. For the most part I follow the published English translation of *Being and Time*. However, I ignore Macquarrie and Robinson's needless, arbitrary, and misleading capitalizing of the word "being." I do this as well, with no further comment, where there are capitalization problems with other words in translation. In German all nouns are capitalized. In English, of course, only proper nouns are capitalized. When moving from German to English, to capitalize a noun interpretively assigns a very substantive "proper noun" tone to the word. In the case of many words capitalized not only by Macquarrie and Robinson but also by other translators of Heidegger this is most often highly misleading for reasons that will become quite apparent in the course of this book.

2. These included "The Origin of the Work of Art" (1935–36), "The Thing" (1951), "Language" (1950–51), "Building Dwelling Thinking" (1952), *What Calls for Thinking* (1951–52), "The Question Concerning Technology" (1953), "A Dialogue on Language" (1953–54), "Memorial Address" (1955), "The Onto-Theological Constitution of Metaphysics" (1956–57), "The Principle of Identity" (1957), "The Nature of Language," (1957–58), "The Way to Language" (1959), "Time and Being" (1962), and "The End of Philosophy and the Task of Thinking" (1964). The dates mark when Heidegger first presented the material to the public, whether in lectures or publications (as far as I can tell).

3. For a longer, more detailed discussion of "thinking the same" see ID 23–41.

4. The other extended discussion is in *Ausgewählte "Probleme" der "Logic"* (GA 45), a book structured much more in traditional academic form.

5. The first and other beginning of Western philosophy as laid out in GA 45 and GA 65 is so fundamental to any discussion of transformation, language, thinking, and our relation to Earth and beings in Heidegger that I find myself either

summarizing it or explaining it in depth over and over. I have already discussed
some of this in similar terms in papers published in 1991, 1993, 1996, and 2001 (see
the selected bibliography).

6. For a thoroughly researched account of what I am only briefly mentioning
here see Merchant, *The Death of Nature*, 164–235. For an equally well researched,
poetically, and brilliantly written account that places the events of the modern era
referred to here in their place in the history of Western philosophy and religion
from the ancient Hebrews and Greeks forward see Griffin, *Woman and Nature*.

7. From an Associated Press wire service report printed in the *Johnson City,
Tennessee, Press*, 13 July 2003: 7A, emphasis added. The quoted economist is Jean-
Paul Moatti.

8. For two interlinked discussions of some these issues see Maly, "Earth-
Thinking"; and Stenstad, "Singing the Earth."

9. Abram, *The Spell of the Sensuous*, 22.

10. Ibid., 56–86. Abram here is inspired not only by his having lived with and
studied oral cultures directly but also by his study of the French philosopher Maurice Merleau-Ponty, who, especially in his last, unfinished work, *The Visible and
the Invisible*, brought forward the core of this thought, that our capacity to think
and speak arises from our inseparability from the flesh of the world, which is
already, before we ever give it a thought, not chaotic but filled with sense (*sens*,
direction and meaning). See also Stenstad, "Merleau-Ponty's Logos."

11. I specify the West here because as far as I know the Sanskrit alphabet arose
quite separately at approximately the same time. I have done no detailed research
on this, but my studies of the Asian religious traditions suggest to me that what
happened on the Indian subcontinent may parallel what Abram outlines in regard
to the Middle East and Greece. As early Hinduism developed, with its ideas recorded
at first in the Vedas and then later the epics and Upanishads, we can see a move
to reify a transcendent, impersonal being—Brahman—from which all other beings
emerge. Brahman is regarded as eternal and unchanging and absolutely real and
in that way is akin to Plato's forms. It differs from them in being undifferentiated,
that is, in having no distinguishing characteristics. It quite simply *is*, and, ultimately,
it is the only thing that truly exists, according to Hindu teachings. This would be
an interesting topic for further study, especially if put into play with the later challenging of this reification by early Buddhism.

12. Abram, *The Spell of the Sensuous*, 107; see also the context at 95–111.

13. Ibid., 253–55.

2. THINKING

1. Hope J. Shand, "Intellectual Property: Enhancing Corporate Monopoly and
Bioserfdom," in Kimbrell, *The Fatal Harvest Reader*, 240–48.

2. Heidegger goes into some depth and detail on this matter, the way that modern and contemporary science works. Though it falls somewhat outside the scope

of this book, what he says well repays careful thinking. To begin working with that see "Modern Science, Metaphysics and Mathematics" (BW 247–82); to pursue it more deeply see GA 65: 141–66/CP 98–115.

3. Curtis White, "The New Censorship," *Harper's Magazine* (August 2003): 15–20.

4. Information taken from Hermann Paul, *Deutsches Wörterbuch* (Tübingen: Max Niemeyer Verlag, 1981), under *sam, sammeln,* and *versammeln.*

5. "What is ownmost" here says *Wesen,* a German word that in metaphysics often means "essence," though it has other meanings in ordinary German, even including "being" in some contexts. *Wesen* as well as *Wesung* in GA 65 are, of course, words of central importance in Heidegger. *Wesung* is included in that list of Ur-rocks earlier in this chapter, but over the years I have become convinced that they (especially *Wesung*) are the most untranslatable of all the key words Heidegger uses. To begin with, there are the many possible meanings of *Wesen,* some of which are readily translatable (sometime *Wesen* really does just mean "essence"). But, of course, as guidewords their multiple meanings are not restricted to the common meanings, and so there has been much contentious discussion about how these words can or should be translated. For one fairly clear discussion of some of the ways they could be translated see Emad and Maly's "Translators' Introduction" to *Contributions* (CP xxiv–xxvii). Unlike them, I am not compelled to attempt to translate the untranslatable. I agree with them, however, that just arbitrarily using the German word in writing about *Wesen* or *Wesung* in English as we do, say, with the word Dasein is not particularly helpful or enlightening. For one thing, Dasein carries fewer possible meanings. For another, and more significantly, it has been in pervasive use in the English-speaking philosophical world for many decades. That is not the case with *Wesen* and *Wesung.* What I am going to do is to say in English what I think is intended by these words without claiming that I am offering a translation of them.

6. Henry David Thoreau, *Walden* (New York: New American Library, 1980), 108.

3. TIMING-SPACING-THINGING

1. "Divinities" is a very easily misunderstood word; a few explanatory words are in order. Divinities here are not God or gods or the god whose death Nietzsche proclaimed (the complex of absolutes around which humans are accustomed to order their individual and social lives). "Divinities" here are closer to the Japanese *kami,* a word often translated as "god" or "divine spirit" but not quite either of those as we in the West think of them. *Kami* is not something apart from things, but neither is *kami in* things (this too would set it apart as something intrinsically apart from the things). Rather, things—rivers, trees, mountains, mirrors, and so on—*are kami.* They are manifestations of the creative arising in which things are in dynamic interplay. This creative arising shines through more powerfully in some things than in others; to these things the name *kami* tends to be more readily applied. The usefulness of the Japanese notion of *kami* in understanding the divinities of

the fourfold is that it helps us keep clear that the divinities are divinities only as they occur in the gathering movement of thinging. For more information on this see LaChapelle, *Earth Wisdom*, 89–91; Jean Herbert, *Shinto: At the Fountain-head of Japan* (New York: Stein and Day, 1963), 21–31; W. G. Aston, *Shinto: The Way of the Gods* (London: Longmans, Green and Co., 1905), 9–11.

2. John Ayto, *Dictionary of Word Origins* (New York: Arcade Publishing, 1990), 491.

3. For more depth and detail see Merchant, *The Death of Nature*, 192–235; Abram, *The Spell of the Sensuous*, 107–22, 130–33.

4. Abram, *The Spell of the Sensuous*, 204–6.

5. Ibid., 207–8. To see the insight from Merleau-Ponty in its context see *The Visible and the Invisible*, 267–68, 130–55. I remember that while I was in graduate school and was enthralled by what I thought were important insights in this unfinished book and the working notes published with it (he had died rather young, in a car accident), I was told that it really wasn't all that philosophically significant, as it was so cryptic and full of "mere" intimations and suggestions and odd language. I ignored all that and kept reading, and I strongly encourage anyone who finds this matter at all interesting to do so as well.

6. Abram, *The Spell of the Sensuous*, 211.

7. Ibid., 212, quoting Heidegger (TB 17).

8. Ibid., 215.

9. Ibid., 216. Some might quibble that Abram's use of "earth" here is different from Heidegger's. Yes, it is broader in what it says, encompassing as it does (I think) much of what Heidegger wants to say by both "earth" and "world" in many contexts. And often, as in "Building Dwelling Thinking," Heidegger has "world" encompassing "earth," whereas I suspect that Abram might reverse this. If I were to have to take sides on this difference in emphasis (which I am not doing at this point, since it is not going to have any significant consequence for what follows), I would go with Abram for reasons that will only gradually emerge. One thing that makes me very cautious in any such "decision," however, is that I am well aware of Heidegger's many warnings against jumping on one instance of usage or even many and then turning them into concepts to be put up for evaluation in philosophical argumentation. Surely, what I have already said in chapter 2 makes clear how misguided such an approach would be. What is important is to follow the movement of showing in the language rather than fixating on particular words or comments as if they were decisive.

10. The one thing worth adding from the otherwise unhelpful attempt to account for our spatiality in *Being and Time* is the clear statement that "space is not in the subject, nor is the world in space" (GA 2: 149/BT 146).

11. Abram, *The Spell of the Sensuous*, 178; Black Elk, *Black Elk Speaks*, 131–39, 255–70.

12. Bokar Rinpoche, *Tara: The Feminine Divine* (San Francisco: Clear Point Press, 1999), 61.

13. In this passage in *Contributions* Heidegger includes a third coupling of space and time: *Zeitraum*. This is not the time-space for thinking here. I omit it from the main discussion because it is not one that would occur to most native English speakers anyway. In German "time" is *Zeit* and "space" is *Raum*. The ordinary German word *Zeitraum*, which in the most literal sense reads "timespace," is actually used in the sense of "span of time," which, as Heidegger points out, is essentially irrelevant to the question in play.

14. Section 242 of *Contributions* is fairly brief, considering what a fully packed and intricate web of thoughts it is. It runs from pages 370 to 388 in the German (GA 65) and from pages 264 to 271 in the translation (CP). Because these few pages are so dense, with many very precise and specific articulations of the matter, and because I want to attempt to clarify them for the reader without generalizing or oversimplifying, I need to temporarily depart from the standard citation form I have been using, namely, giving a parenthetic reference immediately after each quotation or significant new thought that comes from a text. The rest of chapter 3 is essentially a close reading of section 242 of *Contributions*. So from here through to the end of the chapter any quoted material that is not followed by another citation comes from section 242. All other sources, whether from elsewhere in *Contributions* or from other works, will be cited in the usual manner.

15. Merton, *The Way of Chuang Tzu*, 152.

16. Lao Tzu, *Tao Te Ching*, 14.

17. Paul Shih-yi Hsiao, "Heidegger and Our Translation of the *Tao Te Ching*," in Parkes, *Heidegger and Asian Thought*, 93–103.

4. THINKING AS DWELLING

1. The thanc is so important that I very nearly decided to force it on the reader's attention through some graphic device such as all capitals or boldface type. I decided against that for the reason that the thanc is no more strange or remarkable than many of the other guidewords.

2. While in this case I am allowing the translator's arbitrary capitalization of "Being" to remain, it is worth noting that such misleading decisions in the published translations compound the very problem that is the focus of this segment of the dialogue.

3. For thought-provoking, in-depth discussion of this issue I strongly recommend two books. Ehrenreich and English, *For Her Own Good*, gives a well-researched historical account that should make all of us, especially those of us who are women, very cautious indeed about passively accepting expert opinion, whether it comes from the physical, medical, and social sciences or from our educational system without further, very careful investigation. Berry, *The Unsettling of America*, speaks about the devastating results to society, the earth, and human groups and individuals when we passively allow ourselves to be made ignorant and helpless. What he says is particularly relevant to what Heidegger says about dwelling in that Berry links this learned passivity to a context in which (1) we have lost

our own sense of place, of dwelling and caring for our own land, so that (2) those inclined toward exploiting the earth are allowed to run roughshod over those who try to care for and nurture the earth and our fellow living beings, even including our own families. See especially chapters 1 and 2 (3–26).

4. Abram, *The Spell of the Sensuous,* 16.

5. Merleau-Ponty, *The Visible and the Invisible,* 152. See also Stenstad, "Merleau-Ponty's Logos," for an account of Merleau-Ponty's thinking of meaning that brings it into play with Heidegger's thinking. Though a bit sketchy, what I said there still holds up pretty well. See also Abram, "The Perceptual Implications of Gaia."

6. Merleau-Ponty, *The Visible and the Invisible,* 155. See also the entire chapter that culminates in this thought (130–55). A careful consideration of the working notes at the end of the book is also rewarding.

7. Abram, *The Spell of the Sensuous,* 116–17.

8. Ackerman, *A Natural History of the Senses,* 5–6. She makes the point mainly about smell, but, with the two senses so closely linked, it is also relevant to taste, I think.

9. Hazan, *Marcella Cucina,* 5. Hazan is a cooking teacher and cookbook author who *thinks* about food and its preparation. The introduction to *Marcella Cucina* (1–30) says many things that are in accord with what I am saying.

10. Schlosser, *Fast Food Nation,* 120–29.

11. Furthermore, the workers in slaughterhouses and meat-packing plants are among the most brutally exploited of any workers. They are, indeed, treated as no more than units in a gigantic machine (ibid., 169–90). Upton Sinclair would be shocked and disappointed that after all these years so little has actually changed. The situation is probably even worse than when he wrote *The Jungle* back in 1906 due to the speed at which modern meat-processing methods move.

12. Even cooking failures, where something inedible is the result, have their own place in the thanc. Our family lore would not be the same without the tale of the sweet and sour chicken gizzards or the banana meatloaf, which was apparently every bit as weird as it sounds (I was not on hand for that one).

13. The German *achten* would perhaps be translated more usually as "to genuinely heed" rather than "to give our heart and mind to," but in this case I think the published translation *says* just what carries the thinking forward in this moment in *What Is Called Thinking?*

5. The Radiant Emptiness of Be-ing

1. Petzet, *Encounters and Dialogues,* 175–76.

2. Ibid., 180.

3. Others, particularly in the realm of Japanese Zen thinking, have discussed various ways in which the two—Heidegger and Buddhism—have much to say to each other. There are two volumes that come readily to mind. Six of the essays in the relatively early *Heidegger and Asian Thought* bring Heidegger into play with Buddhist thought, with all but one of them having been written by Japanese

philosophers. Stambaugh, *The Finitude of Being*, draws on Mahayana Buddhist thinking—Nāgārjuna's Middle Way philosophy and Dōgen's Zen thinking—to help explain the difference between emptiness and nihilistic nothingness in *Contributions*. Though she conflates finitude with momentary uniqueness, she does open up several relevant paths of thought.

4. Petzet does not say in what year this conversation and television interview took place. From the few indications given in the context, I would guess it may have been sometime between 1960 and 1970.

5. Petzet, *Encounters and Dialogues*, 175.

6. The list of the eighteen endowments of a precious human life can be found in various places. The list I am looking at as I write this is what I wrote down when listening to an oral commentary by Khenchen Palden Sherab Rinpoche (translated by Khenpo Tsewang Dongyal Rinpoche) on Longchen, *The Four-Themed Precious Garland*. This text has much to say about making good use of the eighteen endowments.

7. Khenpo Rinpoche, *Ceaseless Echoes*, 14–15. There are numerous translations of the Heart Sutra in print, most of which differ very little. In addition to the text cited I draw on two other commentaries on the Sutra, both of which include translations of it as well. These are Hanh, *The Heart of Understanding*, and Chang, *The Buddhist Teaching of Totality*, 60–120. All three commentaries are excellent while having differing degrees of accessibility. Thich Nhat Hanh's text intends to make the deep thinking of the Sutra available and understandable to the general reader. Chang's text is a scholarly commentary that assumes some background in Buddhist philosophy. Rinpoche's commentary strikes a balance between these two approaches, giving a precise and scholarly reading of the text that is nonetheless accessible even to those with very little previous understanding of Buddhist philosophy. I should also note that some of what I am going to say is not drawn directly from these three texts but from over twenty years of reading and thinking about the meaning of emptiness in Buddhist philosophy. As the discussion proceeds I will also call on Longchenpa, the great fourteenth-century yogin and philosopher of the Nyingma school of Vajrayana Buddhism.

8. Hinduism is, of course, a very large, complex, and diverse set of teachings and practices. This, however, is a basic idea that is shared by many specific philosophical schools and practice traditions within Hindu culture. It is also a core notion that was rejected by Buddhism as it diverged from its Hindu origins.

9. For a relatively brief historical account of the early Buddhist schools that held various positions on this issue see Dudjom Rinpoche, Jigdrel Yeshe Dorje, *The Nyingma School of Tibetan Buddhism: Its Fundamentals and History* (Boston: Wisdom Publications, 1991), 156–86.

10. Khenpo Rinpoche, *Ceaseless Echoes*, 46, 79–80. See also Longchen Rabjam, *The Precious Treasury of the Way of Abiding*, 96, and Longchen Rabjam, *The Precious Treasury of the Basic Space of Phenomena*, 31, 41.

11. Khenpo Rinpoche, *Ceaseless Echoes*, 58.

12. The Tibetan word *medpa*, which literally means "nonexistent," can also be translated as "ineffable." So in that one word, a word that figures largely in Tibetan Buddhist thinking, we see the convergence of the insight that emptiness is not annihilation but is instead the amazing, mysterious ineffability—inconceivability—of everything that arises and manifests. See the translator's preface to Longchen Rabjam, *The Precious Treasury of the Way of Abiding*, xxii–xxiii.

13. Ibid., 84, 131; and Longchen Rabjam, *The Precious Treasury of the Basic Space of Phenomena*, 9, 19, 27–29, 113.

14. Hanh, *The Heart of Understanding*, 17.

15. Longchen Rabjam, *The Precious Treasury of the Basic Space of Phenomena*, 119, 35, 63, 81, 89, 95; Khenpo Rinpoche, *Ceaseless Echoes*, 42; Longchen Rabjam, *The Precious Treasury of the Way of Abiding*, 79, 86–87, 108.

16. Khenpo Rinpoche, *Ceaseless Echoes*, 43.

17. Kurt Vonnegut, *Cat's Cradle* (New York: Holt, Rinehart and Winston, 1963).

18. Longchenpa, *You Are the Eyes of the World*, 34. See also Longchen Rabjam, *The Precious Treasury of the Way of Abiding*, 77, 149.

19. Longchen Rabjam, *The Precious Treasury of the Way of Abiding*, 109–10, emphasis mine.

20. Ibid., 155.

21. It is worth noting that Vajrayana Buddhism has also thought about language in a way that resonates rather strikingly with "saying" as thought by Heidegger. In a fairly clear instance of this Longchenpa first quotes from a Tantra (scriptural text) that is written as though the creative energy of the universe (timing-spacing-thinging itself) were speaking and then offers a comment. The source text says: "This teacher of teachers, the majestic creative intelligence, displays the integrated structure centered around the inner reality of communication. Everything that exists and is designated displays itself as linguistic communication coming from the unborn field and is gathered into this inexplicable inner reality of communication." Longchenpa goes on to offer this thought on what the Tantra says: "Thus, because all that is present as form, sound, and thought—ever since they appeared in time—has existed as these three unborn integrated structures, from the start live this great natural nonduality without going into any conceptual analysis" (*You Are the Eyes of the World*, 43–44). There is much to be thought here.

6. Staying with Opening

1. Abram indicates that he has had this thought, too, on pages 258–66 of The Spell of the Sensuous.

2. For a fine discussion of this matter of staying with the thinking of Heidegger while also thinking farther see Huntington, "Stealing the Fire of Creativity."

3. Petzet, *Encounters and Dialogues*, 175–76.

4. I base this comment in part on several years of interacting with beginning philosophy students, who generally speak with confidence about the mind-body

distinction but often seem to have trouble understanding the meaning of the subject-object distinction, even though the word "object" is part of their ordinary vocabulary.

5. Someone might say, with some justification, that there is an evolutionary basis for giving body priority over mind, in developmental terms. However, at this point in the thinking presented I am not trying to understand our (prehistorical) development as a species but rather attempting to come to grips with the question of how we might interpret ourselves *now* in the light of the emerging understanding of timing-spacing-thinging.

6. This thought is not original with me by any means. It is one of the central insights put forward by Longchenpa as he tried to bring his experiences of the radiant emptiness of things to language. The inseparability and simultaneity of emptiness (*tong-pa-nyid*) and lucid awareness (*rigpa*) is one of the recurring themes in the work of this brilliant fourteenth-century Tibetan thinker. One clear statement of this point is to be found in Longchenpa, *The Precious Treasury of the Way of Abiding*, 216.

7. I owe this metaphor for the power of conceptual language to continually move as a kind of continual undercurrent and also to resist our attempts to get free of it to Khenpo Tsewang Dongyal Rinpoche in an oral teaching given at Padma Samye Ling, Delaware County, New York, July 21, 2003.

8. The fact that about a quarter of all Americans are functionally illiterate should give us pause as we consider the power of language to constrain or free us. The conceptual "cockroaches," in that case, emerge primarily from the trash cans and refuse heaps of mass media, carefully calculated and constructed by corporate advertising experts and political propagandists. It is quite difficult to imagine the *freeing* power of language at work here.

9. My relatively small knowledge of the Tibetan language is not sufficient for me to know for sure if these words are actually linguistic or etymological relatives, though I suspect that they are. Even if that proves to not be so, bringing them together in this way provokes the opening toward thinking that follows next.

10. Dōgen, *Moon in a Dewdrop*, 101.

11. Ibid., 106.

12. Ibid., 107.

13. Macy, "The Greening of the Self," 57.

14. Heidegger, *Der Feldweg*, 5.

SELECTED BIBLIOGRAPHY

Abram, David. "The Perceptual Implications of Gaia." In *Dharma Gaia: A Harvest of Essays in Buddhism and Ecology.* Edited by Alan Hunt Badiner. Berkeley: Parallax Press, 1990, 75–92.

———. *The Spell of the Sensuous: Perception and Language in a More-than-Human World.* New York: Pantheon Books, 1996.

Ackerman, Diane. *A Natural History of the Senses.* New York: Harper and Row, 1990.

Badiner, Alan Hunt. *Dharma Gaia: A Harvest of Essays in Buddhism and Ecology.* Berkeley: Parallax Press, 1990.

Berry, Wendell. *The Unsettling of America: Culture and Agriculture.* San Francisco: Sierra Club Books, 1986.

Black Elk. *Black Elk Speaks.* As told through John G. Neihardt. Lincoln: University of Nebraska Press, 1988.

Chang, Garma C. C. *The Buddhist Teaching of Totality: The Philosophy of Hua-Yen Buddhism.* University Park: Pennsylvania State University Press, 1971.

Dogen. *Moon in a Dewdrop: Writings of Zen Master Dogen.* Translated by Robert Aitken and others. San Francisco: North Point Press, 1985.

Ehrenreich, Barbara, and Deirdre English. *For Her Own Good: 150 Years of the Experts' Advice to Women.* New York: Doubleday Anchor Books, 1978.

Griffin, Susan. *Woman and Nature: The Roaring inside Her.* New York: Harper and Row, 1978.

Halifax, Joan. "The Third Body: Buddhism, Shamanism, and Deep Ecology." In *Dharma Gaia: A Harvest of Essays in Buddhism and Ecology.* Edited by Alan Hunt Badiner. Berkeley: Parallax Press, 1990, 20–38.

Hanh, Thich Nhat. *The Heart of Understanding: Commentaries on the Prajnaparamita Heart Sutra.* Berkeley: Parallax Press, 1988.

Hazan, Marcella. *Marcella Cucina.* New York: Harper Collins, 1997.

Heidegger, Martin. *Basic Writings*. Edited by David Farrell Krell. New York: Harper and Row, 1977.

———. *Being and Time*. Translated by John Macquarrie and Edward Robinson. New York: Harper and Row, 1962.

———. *Beiträge zur Philosophie (Vom Ereignis)*. *Gesamtausgabe*, vol. 65. Frankfurt am Main: Vittorio Klostermann, 1988.

———. *Contributions to Philosophy (From Enowning)*. Translated by Parvis Emad and Kenneth Maly. Bloomington: Indiana University Press, 1999.

———. *Discourse on Thinking*. Translated by John M. Anderson and E. Hans Freund. New York: Harper and Row, 1966.

———. *Early Greek Thinking: The Dawn of Western Philosophy*. Translated by David Farrell Krell and Frank A. Capuzzi. New York: Harper and Row, 1975.

———. *Der Feldweg*. Frankfurt am Main: Vittorio Klostermann, 1953.

———. *Grundfragen der Philosophie*. *Gesamtausgabe*, vol. 45. Frankfurt am Main: Vittorio Klostermann, 1984.

———. *Holzwege*. *Gesamtausgabe*, vol. 5. Frankfurt am Main: Vittorio Klostermann, 1977.

———. *Identity and Difference*. Translated by Joan Stambaugh. New York: Harper and Row, 1969.

———. *On the Way to Language*. Translated by Peter D. Hertz. New York: Harper and Row, 1971.

———. *Pathmarks*. Edited by William McNeill. Cambridge: Cambridge University Press, 1998.

———. *Poetry, Language, Thought*. Translated by Albert Hofstadter. New York: Harper and Row, 1971.

———. *The Question Concerning Technology and Other Essays*. Translated by William Lovitt. New York: Harper and Row, 1977.

———. *Sein und Zeit*. *Gesamtausgabe*, vol. 2. Frankfurt am Main: Vittorio Klostermann, 1977.

———. *Time and Being*. Translated by Joan Stambaugh. New York: Harper and Row, 1972.

———. *Unterwegs zur Sprache*. *Gesamtausgabe*, vol. 12. Frankfurt am Main: Vittorio Klostermann, 1985.

———. *Vorträge und Aufsätze*. *Gesamtausgabe*, vol. 7. Frankfurt am Main: Vittorio Klostermann, 2000.

———. *Wegmarken*. *Gesamtausgabe*, vol. 9. Frankfurt am Main: Vittorio Klostermann, 1976.

———. *What Is Called Thinking?* Translated by J. Glenn Gray. New York: Harper and Row, 1968.

———. *Was Heißt Denken?* Tübingen: Max Niemeyer, 1954.

Holland, Nancy, and Patricia Huntington, editors. *Feminist Interpretations of Martin Heidegger*. University Park: University of Pennsylvania Press, 2001.

Huntington, Patricia. "Stealing the Fire of Creativity: Heidegger's Challenge to Intellectuals." In *Feminist Interpretations of Martin Heidegger*. Edited by Nancy Holland and Patricia Huntington. University Park: University of Pennsylvania Press, 2001, 351–76.

Khenpo Palden Sherab Rinpoche. *Ceaseless Echoes of the Great Silence: A Commentary on the Heart Sutra Prajnaparamita*. Translated by Khenpo Tsewang Dongyal Rinpoche. Boca Raton, Fla.: Sky Dancer Press, 1994.

Kimbrell, Andrew, editor. *The Fatal Harvest Reader: The Tragedy of Industrial Agriculture*. Washington, D.C.: Island Press, 2002.

LaChapelle, Delores. *Earth Wisdom*. Los Angeles: Guild of Tutors Press, 1978.

Lao Tzu. *Tao Te Ching: The Definitive Edition*. Translated by Jonathan Star. New York: James P. Tarcher/Putnam, 2001.

Longchen Rabjam. *The Precious Treasury of the Basic Space of Phenomena*. Translated by Richard Barron. Junction City, Calif.: Padma Publishing, 2001.

————. *The Precious Treasury of the Way of Abiding*. Translated by Richard Barron. Junction City, Calif.: Padma Publishing, 1998.

Longchen Rabjampa Drime Wozer. *The Four-Themed Precious Garland*. With oral commentary by H. H. Dudjom Rinpoche and Beru Khyentse Rinpoche. Edited and translated by Alexander Berzin. Dharamsala, India: Library of Tibetan Works and Archives, 1979.

Longchenpa. *You Are the Eyes of the World*. Translated by Kennard Lipman and Merrill Peterson. Ithaca, N.Y.: Snow Lion Publications, 2000.

Macy, Joanna. "The Greening of the Self." In *Dharma Gaia: A Harvest of Essays in Buddhism and Ecology*. Edited by Alan Hunt Badiner. Berkeley: Parallax Press, 1990, 53–63.

Maly, Kenneth. "Earth-Thinking and Transformation." In *Heidegger and the Earth: Essays in Environmental Philosophy*. Edited by LaDelle McWhorter. Kirksville, Mo.: Thomas Jefferson University Press, 1992, 53–68.

McWhorter, LaDelle, editor. *Heidegger and the Earth: Essays in Environmental Philosophy*. Kirksville, Mo.: Thomas Jefferson University Press, 1992.

Merchant, Carolyn. *The Death of Nature: Women, Ecology and the Scientific Revolution*. San Francisco: Harper and Row, 1980.

Merleau-Ponty, Maurice. *The Visible and the Invisible*. Edited by Claude LeFort. Translated by Alphonso Lingis. Evanston, Ill.: Northwestern University Press, 1968.

Merton, Thomas. *The Way of Chuang Tzu*. New York: New Directions, 1965.

Parkes, Graham, editor. *Heidegger and Asian Thought*. Honolulu: University of Hawaii Press, 1987.

Petzet, Heinrich Wiegand. *Encounters and Dialogues with Martin Heidegger*. Translated by Parvis Emad and Kenneth Maly. Chicago: University of Chicago Press, 1993.

Schlosser, Eric. *Fast Food Nation: The Dark Side of the All-American Meal*. Boston: Houghton Mifflin, 2001.

Stambaugh, Joan. *The Finitude of Being.* Albany: State University of New York Press, 1992.

Stenstad, Gail. "Attuning and Transformation." *Heidegger Studies* 7 (1991): 75–88.

———. "The Last God: A Reading." *Research in Phenomenology* 23 (1993): 172–83.

———. "Merleau-Ponty's Logos." *Philosophy Today* (Spring 1993): 52–61.

———. "Revolutionary Thinking." In *Feminist Interpretations of Heidegger.* Edited by Nancy Holland and Patricia Huntington. University Park: University of Pennsylvania Press, 2001, 334–50.

———. "Singing the Earth." In *Heidegger and the Earth: Essays in Environmental Philosophy.* Edited by LaDelle McWhorter. Kirksville, Mo.: Thomas Jefferson University Press, 1992, 69–75.

———. "The Turning in *Ereignis* and Transformation of Thinking." *Heidegger Studies* 12 (1996): 83–94.

INDEX

abandonment of being *(Seinsverslassenheit)*, 19, 21–22, 23–24, 25–26, 52–53, 148, 177

ab-ground *(ab-grund)*, 19, 28, 103, 106, 111; as abyss *(abgrund)*, 150; being as, 59; be-ing as, 82, 110–11, 150, 151; be-ing as saying, 130; emptiness *(Leere)*, 110–11; as ground, 149

Abram, David, 38–39, 95–97, 137–38

abyss, 150

acceleration, 27, 56, 172–73

achten, 212n13

active-passive dichotomy, 141

AIDS, calculative thinking and economic costs of, 23

aletheia (truth), 18, 24, 27, 62–63, 153

alphabetic writing, 38–39, 84, 94–95, 173

ambiguity, 23, 27, 186

anatman, 163–64

animals, 36–37; encounter with, 4, 9, 132–33; language and, 97–98; "rational animal" concept, 192; as things, 136–37; time and, 97

answers: linear thought processes and definitive answers, 61, 90; as

transformation rather than solutions, 61–62, 64

appropriation. *See* enowning *(Ereignis)*

arche, 175, 186–87, 188

arising, 105; of being, 24, 60; as central in *Being and Time*, 24; interdependent arising or co-originating, 165; of language, 75–76; of *physis*, 20, 22, 24, 59–60, 94–95, 147; *pratitya samutpada* (interdependent arising), 73; as source of wonder, 100; as speaking, 75; of time, 85; *Wesen and Wesung*, 103, 109n5

Aristotle and Aristotelian thought, 18, 20, 46, 60, 65, 94, 182–83

attuning, 16, 81; *Befindlichkeit* (finding oneself attuned), 30–31; to beginning, 52; as preparation, 53–54; of questioning, 61–62; reservedness as grounding attuning, 62, 69, 76, 115, 127; wonder and, 26, 52–53

authenticity, 176–77, 183–84

Bacon, Francis, 20, 46

Basic Writings, 11